Richard Touitou

The Philosopher's Stone
Or
The Great Magisterium

Symbolism and Operating Processes
of the Great Work

Copyright © 2024 - Richard Touitou

WARNING

In the operations described in this study, I have endeavored to provide all necessary precautionary advice to any amateur wishing to proceed to the experimental stage to avoid an accident.

The operations require knowledge of chemistry and adequate training in metallurgy.

Recommendations for moving on to the experimental stage:
1 - Wear safety goggles
2 - Wear fireproof gloves
3 - Use special tongs to handle the crucibles
4 - Use only good-quality, clean crucibles
5 - Open the windows when using the furnace, or better yet, use a fume cupboard.

The risks are threefold:
– Poisoning by gases emitted during calcination
– Crucible explosion
– Causticity of the regulus due to the formation of potash.

I must therefore decline all liability in the event of an accident.

CONTENT

Page	
5	Preamble
15	Return to Origins
23	Alchemical Chronology
28	Alchemy and Religions
32	Alchimy and Spagyria
33	Alchemists or Blowers?
38	Chronology of Some Transmutations
41	Esotericism and Initiation in the 20 - 21st Century
48	Birth and evolution of a theory
57	Alchemical Glossary
58	Some definitions to remember
59	Manufacturing Secrets
60	Warning
62	The subject of Art - The materia prima
70	Mars, the Vaillant Knight - the giver of Sulfure
72	The Mediator, the Third Agent, the Salt, the Secret Fire, the Water of the Wises
79	The Dew
82	The Philosopher's Stone in 5 points
88	Dry Way Process
92	Collection of dew
98	First Work
99	The practice
108	Operations of the First Work.
112	Second Work
117	Purification
121	Eagles or sublimations
124	Coction - Multiplication - Projection
127	Practical considerations
133	Transmutations
139	The Portal of Philosophy
147	Fontaine de Mars
154	Addendum - The Wet Way
158	Bibliographic index

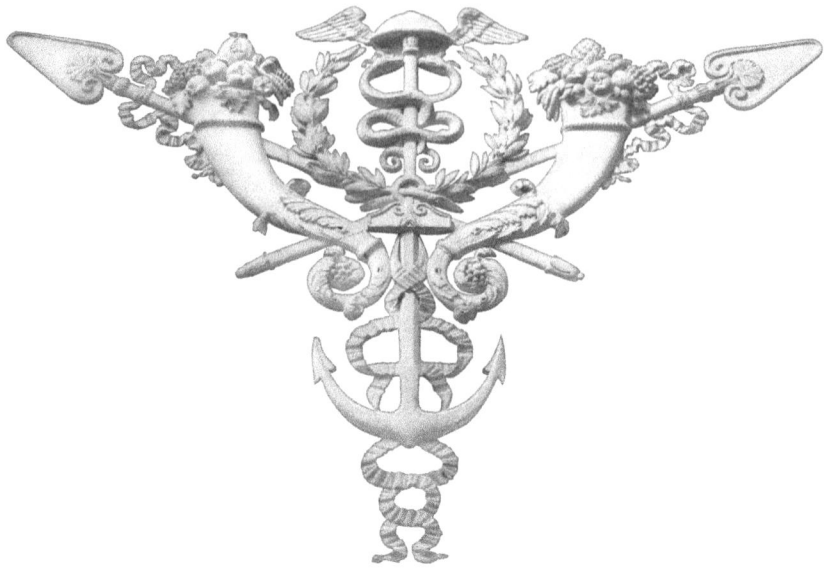

Paradigm of the Great Magisterium

This fresco is reproduced several times
on the walls of the Galerie Vivienne in Paris.

At the top, the caduceus of Hermes,
the Mercurial emblem, recognizable by the two snakes
which intertwine a rod topped with a winged helmet,
symbol of volatility, and placed in the center of a laurel wreath,
the Caput Mortuum.

At the bottom a marine anchor, emblem of the fixed corps.
In the center, two intersecting spears represent the crucible,
the broad points decorated with the merelle dear to Fulcanelli,
symbol of the receptacle of the Spirit.

Two cornucopias represent the promised reward.
The symbolism shown is identical to that in the illustration
from the treatise of Chevalier Picolpassy where we see a heavy stone
tied to the foot of a white dove trying to take flight.

Just like the white bird,
the Mercury must raise the sulfure above the compost.
Almost impossible mission
for those who don't know *"the good stuff"*.

Preamble

My very first contact with alchemy dates back to the early sixties. I was then a student at the Geneva chemistry school.

One day my professor asked me to give as complete a presentation as possible on the history of chemistry.

To carry out this study, I easily managed to bring together several interesting books on this vast subject, our university library was very well stocked. Luckily, modern authors, such as Serge Hutin[1] and M. Caron[2] as well as Mircea Éliade2, had published fascinating works in the 1950s, which allowed me to . A century earlier, the famous chemist Marcelin Berthelot[3] had undertaken remarkable research work resulting from the study of old documents and treatises prior to the 8th century, the Leyden papyri, the manuscripts of Saint Mark, as well as old manuscripts from an Italian museum. His "*History of Chemistry*" can be considered a monument of scholarship.

I recognize that my predecessors had given me considerable help in writing this memoir covering more than thirty centuries of history, and my presentation captivated all the attention of the amphitheater.

I had just found a new occupation that would take up all my free time for decades to come.

I remember feeling great frustration following this first contact with alchemy. Questions raced through my mind, excited by an indescribable and inexplicable curiosity.

Who were these alchemists? Were they subtle forgers, charlatans, or scholars? What was this mysterious process of making the Philosopher's Stone and what basic materials did the alchemists use? On what theoretical foundations were the Adepts based? Were there successful weight transmutations over the centuries? Why did you hide all the secrets accumulated over two millennia of research?

All these questions seemed like they could never be answered, as this mysterious universe seemed inaccessible and completely anachronistic. My curiosity was piqued, so I had to discover at all costs what was hidden behind these monuments of enigmas, symbols and secret allegories, supposedly reserved for a small caste of initiates or enlightened people.

I made a promise to myself to take up this challenge in order to dismantle once and for all the immense deception that constituted the search for the absolute, and to reveal the true facets of hermeticism in broad daylight.

I had obviously underestimated this task.

It took nearly four decades of investigation. Forty years at the end of which I had to store a large mass of data that I had to classify, study, dissect, analyze and cross-check, to finally extract its quintessence.

This required comparing discoveries with the current possibilities of chemistry, metallurgy and physics, learning French which had not been used for centuries and overcoming the linguistic difficulties of Latin and ancient Greek.

This was not enough and I had to memorize dozens of works, decipher incredible enigmas, rebuses, parables or allegories, to avoid the traps and fallacious recipes intentionally created to confuse lay people.

Here, in a few lines, are the main findings resulting from this unusual research for a modern scientist and which will be discussed in detail later:

– there is a significant number of alchemical writings which have been the subject of hundreds of publications over the last ten centuries as well as an even more considerable number of unpublished works. The majority of these works lie dormant on the shelves of large libraries or those of private collectors, waiting for a lucky artist to finally decide to decipher their contents.

– alchemy books can take on monumental appearances and this is not an image. The ancient Adepts, to transmit the secrets of the Great

Art, built cathedrals, mansions, created statues, tapestries or paintings, engravings and numerous objects dedicated especially to alchemy. They had amulets engraved and sculpted fireplaces and stoves. Gold or silver coins and medals were often minted on the occasion of successful transmutations. Fulcanelli has masterfully demonstrated that all these masterpieces are the safest and richest "books" in teaching of the entire alchemical corpus.

– over the last twenty centuries, the alchemical current, stronger than the American gold rush, has affected all segments of the population, men and women of all faiths; rabbis, monks, bishops and popes have practiced there, as well as poets or writers, painters or sculptors, famous doctors, lawyers, princes and princesses, kings and queens, bourgeois, peasants, blacksmiths, gentleman glassmakers, goldsmiths, architects without forgetting the scientists, of course.

– over a period covering the last ten centuries, we see that there is perfect continuity in the definition of the basic principles, materials and the alchemical process although there are several schools (meaning several paths). Today there are at least four:
- *the regulin route or royal route* (metal route)
- *the wet route*
- *the ancient way* (animal way)
- *the natural way* or plant way
- *the cinnabar route*, another metallic route, of Sino-Hindi origin.

This study only concerns the royal rout which, in theory, involves at least two distinct development processes; one wet and long, the second dry and short. Know that the acceleration of a chemical process when the reaction temperature increases is a scientifically established fact. Roughly speaking, the speed of a chemical reaction doubles approximatly with every 10 °C increase. One of the main alchemical claims is therefore proven. There may indeed very well exist two operational processes of the same reaction, or of the same group of reactions, one slow, carried out at moderate temperature, the other very rapid, involving much higher temperatures.

– the alchemical process(es) have never stopped evolving and improving over the centuries, either to increase the transmutatory power of the Stone, or to reduce its elaboration time.

– for each of the two regulin pathways, there are several possible variants, provided that we can concentrate the "universal spirit" within the philosophical compost.

– flowing mercury, (Hg), is not the Prima Materia of the Great Work, although some Adepts, like Flamel, Philalethes, Urbiger or Cyliani implement it at a certain stage of the process where it would only intervene as a simple adjuvant, the *"fugitive servant"*. Some modern authors like Caro or Duschene have worked on this path and published numerous works and claim to have obtained concrete results. I will refrain from publishing any comments on these processes because I have not personally studied them.

– Fulcanelli, Eugène Canseliet and many other contemporary alchemists apparently knew nothing about the wet process often adorned with the *fallacious* qualifier.

– The phenomena described in the first alchemical preparations have been known and explained by chemists for a long time, but those of the second and third regimes completely escape any scientific explanation, they seem to be a miracle.

– The allegorical figures can only be understood by those who work in the stove, they are the pictorial reflection of the observable phenomena of cooking.

– There exist, alongside the canonical alchemical process, spagyric or archchemical processes leading to ponderable transmutations. They have been the subject of several publications, the most complete being that of B. Husson[4].

– Most of the so-called *"special"* processes use processes derived from the Great Work and on which they bring *"the light of nature"* into play.

– Regarding the chemicals used by the Adepts in the metallic path, two elements belong to the same family in Mandeleev's classification: nitrogen, combined in one of the saline adjuvants, and antimony, the secret raw material. Are they interchangeable? Is this a coincidence?

Azoth and fire are enough for you, says the adage. Nitrogen undoubtedly conceals the most secret arcana of the Great Work.

– Dry alchemical cooking in a crucible gives the compost an incredibly high density and the property, for the Stone, of being improved in quality and quantity, infinitely. The final concoction is marked by a sound manifestation punctuating the increase in weight and density, a sure indicator of success.

– The final phase of multiplication gradually modifies the physical properties of the Stone which tends to no longer coagulate and to leave the solid state to become liquid, showing increased phosphorescence. The transition from the solid state to the liquid state suggests an ultimate transformation towards the gaseous state if the multiplications are continued...

You have just glimpsed some of the know-how of our predecessors as well as the main object of their concerns.

The Adepts claimed to be able to create a chemical entity – the Philosopher's Stone – a material with properties unknown to modern chemistry and physics. A synthetic material, which has the appearance of crushed glass, garnet colored in its solid form, very fusible and which also has an incredibly high density. It would act as a catalyst capable of causing transmutations on molten metallic bodies. Other properties are attributed to it such as curing many illnesses and prolonging life.

What fascination has produced the simple evocation of the word *"alchemy"* over the past centuries, beyond any philosophical, mystical, paranormal or magical consideration?

Should we emphasize that transmutation is an incontestable physical reality? It is she who makes nuclear power plants operate and, unfortunately, atomic bombs explode. Transmutation is a natural property of radioactive material which degrades according to a process linked to a specific probability of the element, which process still escapes any rational explanation today.

Science postulates that massive transmutations cannot be caused without involving colossal energies, but these arguments are not very convincing and particle accelerators have not yet exhausted all of their resources. The mysteries surrounding the atom are therefore far from being resolved.

Today we are entering a new era where physics desperately tries to explain the atom using quantum mechanics, and the universe with the theory of Relativity. Scientists are confronted with a dimensional problem, concerning two diametrically opposed universes: on one side the infinitely small, on the other, the infinitely large.

In microcosm, basic research is a very efficient demolition business. We break up matter to explain what it is, in increasingly larger and more powerful colliders. Is this the right method?

In the bubble chambers, it is said, twenty-four elementary particles have been found, which are only the residue of collisions, debris of atoms which we suppose to be the basic constituents. It's a bit light. To validate these results, to prove the theories developed over half a century, it will be necessary to experimentally prove the reversibility of the observed phenomena, that is to say reconstitute the destroyed atoms using their debris. It's a difficult bet to keep.

In the same spirit, with string theory currently being established, we have created a purely mathematical being to explain the atom or the particle: a vibrating string of infinitesimal length, the Plank length. The most annoying thing is that there is no experimental counterpart yet and this relegates this theory to the rank of simple curiosities, after more than two decades of research and investigations.

Finally, current research carried out on vacuum energy, dark matter, antimatter and cold fusion, if successful, will undoubtedly be able to accredit the empirical theories of ancient Philosophers.

Alchemy is a discipline closer to solid state physics, metallurgy and... cosmology, than to chemistry. The basic principles are based on the use of a limited number of materials which, under certain conditions, are capable of capturing and accumulating an unknown form of energy, a bit like the way an electric capacitor charges, but while retaining its specific, indeterminate character. It is this energy which will allow the Philosopher's Stone to carry out transmutations.

The physical processes implemented by alchemists have the possibility of profoundly modifying the structure of the atom. The non-oriented spray powder resulting from these processes would also have surprising therapeutic properties.

What are these materials and what is this unknown energy? What techniques did alchemists develop to achieve the Great Work?

The answers to these fundamental questions have been scattered throughout the immense alchemical corpus that the Adepts have left us. The Adepts preferred to conceal their secrets in messages intended only for other initiates, their peers. Thus, like magicians, they give us an overview of their know-how, making us believe that their writings are accessible to all and that they will facilitate the development of the Great Work.

Make no mistake, without initiation, no one would be able to correctly untangle these tangles. Fortunately, some Masters of the Art, following the imposing legacy left by Fulcanelli, have had the courtesy to lay down some milestones

---

1. M. Caron et S. Hutin, Les Alchimistes (*The Alchemists*), Paris Édi.du Seuil, 1959.
2. Mircea Heliade, *The Forge and the Crucible.*, 1956.
3. Marcelin Berthelot , Les Origines de l'Alchimie (*The Origins of Alchemy*), G.Steinheil 1885.
4. Bernard HUSSON, Transmutations Alchimiques (*Alchemical Transmutations*). Paris. Éd. J'ai lu, 1974.

THE PHILOSOPHER'S STONE

I support the thesis according to which science,
like any form of thought,
is just an organized system of beliefs.
Dr. Jean-Pierre Petit - Astrophysicist

Return to Origins

Numerous attempts to locate the origins of alchemy in space and time have been made over the past three centuries. Among the most serious we note in chronological order, the study of Abbé Lenglet-Dufresnoys[1] in 1742, of Marcelin Berthelot[2] in 1885, of Serge Hutin and M. Caron[3] in 1951 and 1959 respectively, as well as that of the philosopher-historian Mircea Heliade[4].

It seems appropriate to correct certain gross errors, tirelessly repeated, often because of preconceived ideas, but sometimes resulting, no doubt, from superficial studies.

Above all, we must exclude from this historical research any consideration relating to Chinese and Indian alchemy, which is too recent and very far from the European Hermetic tradition, both at the theoretical and practical level.

Most of the time, the majority of Hermeticists and a good number of historians, Berthelot and Fulcanelli excepted, have attributed Egyptian or Greek origins to alchemy, this probably due to the irruption, very early in beliefs, of the mysterious multi-faceted character that was Hermes, more divinity than Mesopotamian dignitary or priest, exiled, according to legend, in Egypt. Furthermore, the countless references to Plato, Aristotle or Democritus made by several Arab or Christian authors have contributed significantly to spreading this frequent error which consists of locating the cradle of alchemy in Greece.

Marcelin Berthelot, in his *"History of Chemistry"* presents a very critical analysis of these misunderstandings:

"If the alchemists were attached to Hermes, if they dedicated mercury to him, the raw material of the Great Work, it is because Hermes,

in other words Toth, was reputed to be the inventor of all the arts and all the sciences. Plato already speaks of it in his dialogues, such as the Philebes and the Phaedo. Diodorus of Sicily dates back to Hermes the invention of language, writing, the worship of the gods and that of music; likewise the discovery of metals, that of gold, silver, iron in particular. Hermes seems to have personified the science of the Egyptian priesthood. He was the Lord of divine words, the Lord of sacred writings."

Let us point out in passing that Berthelot neglects to specify here that the mythical Hermes is not Egyptian but... Babylonian.

Thus, contrary to popular belief, there could never have been a typically Egyptian alchemical tradition. Without calling into question Egyptian genius, it should be noted that the secrets *"jealously"* guarded by the priests of the pharaohs only concerned metallurgical processes relating to the working of gold and especially its falsification and doubling, as well as manufacturing recipes. dyes, enamels and chemical bases for embalming techniques. Some of these techniques are also clearly displayed on tomb frescoes.

By delving deeper into the history of Sumerian civilizations, another hypothesis concerning the origin of alchemy appears more plausible, the Mesopotamian sector. The clay tablets of Nineveh teach us that the Sumerians cultivated sciences at least four thousand years before our era. They mastered mathematics, astronomy, metallurgy, chemistry and especially writing used to transmit knowledge. The Hebrews, originally from Mesopotamia, emigrating to Egypt 2000 years BC, imported this culture of the highest scientific level for the time, as well as the techniques of the arts of fire whose transmission was carried out from father to son; in short, everything that will allow us to weave the fabric of this alchemy currently spreading in the Mediterranean basin.

The first chemical operations to purify gold, such as cupellation for example, were used at least 3000 years before our era, in the northeast of Asia Minor, by the Babylonians. We find precise allusions to the refining of gold in the Old Testament, as well as a very ancient Jewish treatise, written in Hebrew, *"the Aesh Mesareph"* or the Refining Fire.

It is also in the Old Testament that the first references to iron and its metallurgy appear, another fundamental element in the development of the Philosopher's Stone.

Finally, it should be noted that no papyrus or hieroglyph relating to real Egyptian alchemy before Ptolemy and Cleopatra has been discovered to date.

Berthelot made an exhaustive study of the Leyden papyri, discovered in a tomb in Thebes, and he evokes the Jewish network with conviction:

"Thus, he writes, in papyrus no. 75 of Reuvens there is an alchemical recipe, attributed to Hosea, king of Israel. In other papyri from the same family, we read the names of Abraham, Isaac, Jacob, the word Sabaoth and several other passages relating to the Jews."

"Papyrus No. 76 contains a magical and astrological work, entitled: the Holy Book, called the eighth Monad of Moses, the key of Moses, the secret book of Moses. The names and Jewish memories are therefore mixed with occult sciences, at the time of the first writings alchemical, that is to say around the 3rd century AD."

In ancient Egypt only recipes for falsification of gold and silver flourished. The Thebes Manuscript is also the first document that describes the ways of imitating or increasing gold and silver as well as the art of counterfeiting precious stones.

The forgers, with the arts of fire, had learned, empirically, that the external appearance of matter could be modified at will, leaving an important part to superstition to explain what remained mysterious.

Glass was made from sand, lead from galena and so on using relatively simple techniques, using reagents that were easy to acquire and commonplace at that time. Thus, it was thought that, to transform lead into gold, for example, it would be enough to find a way to increase its density and change its color using an appropriate reagent or additive.

But these were in no way alchemical techniques.

Around 340 BC, Aristotle, disciple of Plato, developed his physics of the quality and quaternary of the elements as well as principles on the evolution of matter. With the diffusion of these revolutionary theories, the era of the Neoplatonists was permanently established.

Opportunistic counterfeiters then began to explore the behavior of the material from every angle and exploit its properties as best as possible, with the aim of creating imitations of gold and silver. To achieve their goal they achieved a complete synthesis of the arts of fire and soon ended up becoming very good generalists, perhaps the first true chemists in the history of science.

The apprentice forgers had become scientists, in a way, not hesitating to embark on hazardous operations, to handle toxic or explosive products, to manufacture alloys having the appearance of precious metals. Fraud had become commonplace because detailed analysis procedures could only be carried out in workshops equipped with very sophisticated equipment. At the same time, all the bases of our chemistry were gradually established.

In the 2nd century BC, Archimedes developed a simple way to detect adulterated gold alloys: double weighing, the first revolutionary method of rapid non-destructive analysis.

The alchemy which continued to develop in the meantime, however, had little to do with the generations of forgers mentioned above. The principles to which it relates are more noble, they go beyond the simple scope of the lacewing, transmutation being only the final index of success.

This science, which requires the mastery of very elaborate techniques, could not have arisen spontaneously, like that of glass production, for example. It took many centuries of trial and error, which continued until the Renaissance, for it to reach its peak.

Three conditions had to be met to allow the emergence of alchemy:
– perfect mastery of the arts of fire, ovens and pyrotechnic equipment,
– knowledge of mining and metallurgical extraction techniques, as well as ore enrichment techniques,
– knowledge of the materials necessary to accomplish the Great Work, (chemical agents included), and that of their chemical-physical

properties. By going back in time, it will therefore be possible for us to establish a precise calendar:

– the discoveries of antimony and copper date back to 4500 BC.
– gold purification techniques were developed around 3000 BC.
– iron is essential to the alchemical process, the first meteoric iron objects date from 1900 BC. while the Iron Age, strictly speaking, only dates from the 13th or 14th century BC.

We can therefore estimate that several centuries must have passed since the beginning of the Iron Age, before alchemical techniques were refined and reached maturity, which amounts to saying that embryonic alchemy did not could have seen the light of day only twelve or thirteen centuries BC.

In the biblical Exodus, Berthelot noted this very revealing passage about the metallurgical techniques used by the Hebrews: "And the Lord said to Moses: I have chosen Beseleel, priest of the tribe of Judah, to work gold, silver, copper, iron and everything that concerns stones and woodwork, and to be the master of all the arts..."

"...Then come a series of purely practical recipes, placed under the patronage of Moses and Beseleel. We know that the latter is given in Exodus as one of the builders of the Ark and the Tabernacle. There is a rabbinic connection in all this, and a first indication of the sources and secret doctrines of Freemasonry in the Middle Ages."

This last remark relating to Freemasonry is very relevant, Freemasonry and alchemy have always been closely linked.

Who does not know this delicious piece of poetry attributed to King Solomon, the Song of Songs? This is the first allegorical alchemical poem that Fulcanelli has reported. (The legendary mines of King Solomon having never been discovered, the temptation would be strong to attribute an alchemical origin to Solomon's wealth).

Then there is a large, centuries-long hole in our timeline. Few writings, no papyri or parchments. Were the alchemical manuscripts of this period all destroyed during the fire which ravaged the great library of Alexandria, during the reign of Caesar? This is very likely, but fortunately, an oral tradition made it possible to safeguard this heritage until the Aristotelian period, when the theory of the four elements emerged.

Who then will help to spread the new alchemical ideas from this time on?

The Jews, once again, according to Marcelin Berthelot: *"The role attributed to the Jews for the propagation of alchemical ideas, recalls that which they played in Alexandria, during the contact between Greek culture and Egyptian and Chaldean culture. We know that the Jews have a primary importance in this fusion of religious and scientific doctrines of the East and Greece, which presided over the birth of Christianity. The Alexandrian Jews were for a time at the head of science and philosophy."*

So, we find the Jews again, just as alchemy takes the final turn, some 2300 years or so. We must not forget that the first alchemical writings that have reached us only date back to the 2nd century AD, while forgery recipes circulated three or four centuries BC.

An ancient manuscript, the Book of Mysteries, attributed to a certain Apollonius of Tyana, circulated from hand to hand until at least the 10th century; it is among the most hermetic treatises in the history of alchemy. Apollonius of Tyana would be a fictional character according to Fucanelli who thinks that, under this pseudonym, lies the secret of manufacturing philosophical mercury (Apollonius referring to the Sun and Tyana, that is to say Diana, to the Moon). In any case, Arab authors also attribute to this controversial character the writing of the Emerald Tablet, the most esoteric, the most edited and the most commented on of the condensed works of the Great Work.

Arab alchemists placed several Greek philosophers among their followers, such as Pythagoras, Archelous, Socrates, Olympiodorus, Plato, Aristotle, Porphyry, Galen as well as a mysterious Bolos known as the Democrat of Mendes. They wrongly considered them as precursors.

But it is important to note that the first known practical alchemy manuscript is attributed to Myriam (*Marie*) the Jew who lived around the 3rd century, in Alexandria.

Myriam invented the famous "*Marie bath*", gave the description of numerous laboratory devices, some of which are still used today, sublimatory vases, stills and athanors.

Regarding athanor, it is necessary to correct a frequent error relating to the origin of this word which comes from Hebrew and not from Arabic: ha-thanour literally means *"the oven"*. In Arabic *"al tanour"* undoubtedly derives from Hebrew. Publishers of French language dictionaries who systematically reproduce this type of error should know this. We must not forget that Hebrew predates Arabic by almost a millennium and that the Arabs borrowed many words from ancient Hebrew. The ternary root of words in the Arabic language is also modeled on that of Hebrew, as is the absence of vowels in the alphabet. This parenthesis having been closed, we also note that many words borrowed from Arabic have remained in our everyday language; for example the still (*al inbic*) or alcohol (*al kohol*). We will see later that the word *"alcohol"* conceals one of the major mysteries of the Great Work.

Let's return to Myriam the Jewess who is cited in the El Firhist encyclopedia of the Arabic Al Nadim. El Firhist reveals that Myriam would have introduced Zozimus the Panopolitan to alchemy, despite the warning she gave him:

"How dare you touch the Philosophers' Stone, you who are not of the stock of Abraham?"

This revelation makes it possible to rectify a major historical error, that of making Zozime the initiator of Myriam.

Myriam, who enjoyed excellent credit among Arab philosophers, was considered a disciple or daughter of Plato by Al Nadim, thus marking his esteem for the Jewish alchemist, she would therefore be the author of the very first true alchemical manuscript. , a true alchemical manuscript as opposed to the manuscripts of the authors of forgery recipes, magic or witchcraft which I exclude from this study.

From the 4th century, the School of Alexandria, with Zozimus as initiator, his disciples, Pseudo-Democritus, then Synesius as well as Olympodiore, gained momentum in the Mediterranean basin. Many alchemical laboratories were operating at this time throughout the Nile Delta. We will find traces of numerous Alexandrian manuscripts also cited in the Al Nadim encyclopedia.

Then alchemy began to spread in the Christian world just after the period of the Judeo-Christian schism, that is to say around 325 AD.

Then, towards the end of the 4th century, the fanaticism of the Christian monks raised by Archbishop Theophilus will cause the annihilation of the School and the Library of Alexandria, as well as the destruction of most of the Alexandrian chemical laboratories, that of the magnificent Serapeum of Memphis included.

It was a very great loss for civilization.

Around the 7th century, Morien, disciple of an alchemist named Adfar, taught the sacred art to the Arab prince Chalid ibn Jazid, and from then on, according to Serge Hutin, alchemy was cultivated among the mystical communities of Islam and especially by the Sufis.

The great Muslim invasion of North Africa and Spain will certainly promote the alchemical movement in Europe. Great figures like Chalid, Rhases and Avéroes left a deep mark in the Middle Ages, as evidenced by numerous texts. But not all of these characters can be placed among the followers.

Rhases (Abu Bakr Muhammad ibn Zakariyya al-Razi), like Avicenna, was a renowned doctor.

Avicenna did not believe the transmutations were possible.

Finally Averoés was a true humanist philosopher in all the modern meaning of the term. It should be noted that he also maintained privileged relations with the great rabbi Maimonides.

Artéphuis, alias Al Tohgra'i, is certainly the most learned of all Arab alchemists, and perhaps even the greatest adept of the Middle Ages. His manuscript, "The Secret Book of the Very Ancient Philosopher Artephius" is a modern treatise of great didactic value, written in barely veiled language. The Arab author would have lived in the twelfth or thirteenth century.

Concerning the great Geber, Jabir ibn Haiyan, his *"Sum of Perfection"* appears as a work of chemistry or spagyrics, written in Latin… two to three centuries after his death. No original written in Arabic has ever been found.

It is therefore clear that there were few Arab Followers apart from Artephuis, Calid or Rases. In truth, there are few interesting Arabic alchemical texts, apart from the usual treatises on magic or mineral spagyrics.

Could there have been a secret Arabic alchemical corpus of which we have lost track, due to lack of translators or an informed readership?

The transfer of know-how to Christians seems rather to have taken place in a manner entirely independent of the Islamic invasion, between the 9th and 12th centuries (it shortly preceded the explosion of Gothic Art in Europe), just at the time of the Crusades.

What authority was able to transmit the imposing know-how to the crusaders in Jerusalem? Where did the knowledge come from that allowed the construction and proliferation of cathedrals between the twelfth and fourteenth centuries in Europe, as well as the barely coded alchemical messages transcribed on stained glass windows, statues and frescoes on the great portals of cathedrals of Paris, Amiens and Chartres or Toledo ? The explanations of Fulcanelli[5] who sees the imprint of Catholicism and the Christian religion in everything relating to alchemy are too partisan and irrational. The secrets of these transfers are perhaps hidden in some obscure cellar in the Vatican.

From the twelfth century onwards, European alchemy was very quickly structured to take on a new dimension until the Renaissance.

Alchemy has never been fixed, immutable as some maintain. On the contrary, hermetic processes will be refined as alchemical treatises spread throughout Europe with the invention of printing and movable type. From now on, alchemy will be cultivated in all strains of the population, and renowned scientists will mark this era: Jehan de Mehung, Arnaud de Villeneuve, Nicolas Flamel, in France, Basile Valentin and Henri Khunrath in Germany, Ripley, Norton and Crémer in England, and then Sethon alias the Cosmopolite, Philalethes. At the same time, Adepts, or their emissaries, are making constant efforts in European capitals to demonstrate the effectiveness of their projection powders and the alchemical reality. Testimonies abound and numerous medals are struck to commemorate public or private transmutations.

The decline of alchemy began in the 18th century, with the era of the great chemists. There will be very few initiates left and even fewer Adepts like Cyliani in the 19th century and Fulcanelli in the first quarter of the 20th century.

Fortunately, great hermeticists like René Alleau, Claude d'Ygé, Eugène Canseliet, Bernard Husson and Atorène have left us a precious legacy of very openly veiled philosophical works, a rare legacy of this importance in the entire history of alchemy.

Today, in the era of space conquest, the supporters of alchemy have not said their last word. In France alone, there are several thousand "art lovers", including a few hundred alchemists working secretly in their cellars.

These new "artists" are very modern. Their athanors run on gas or electricity, their ovens are equipped with electronic regulators and temperature sensors, and they are now capable of analyzing and understanding the mechanisms they use in the operations they trigger.

A phenomenon of civilization, the transfer of information is now done via the Internet and there are sites on which one can download countless and precious alchemical works, very little previously distributed, and some forums where ideas are exchanged.

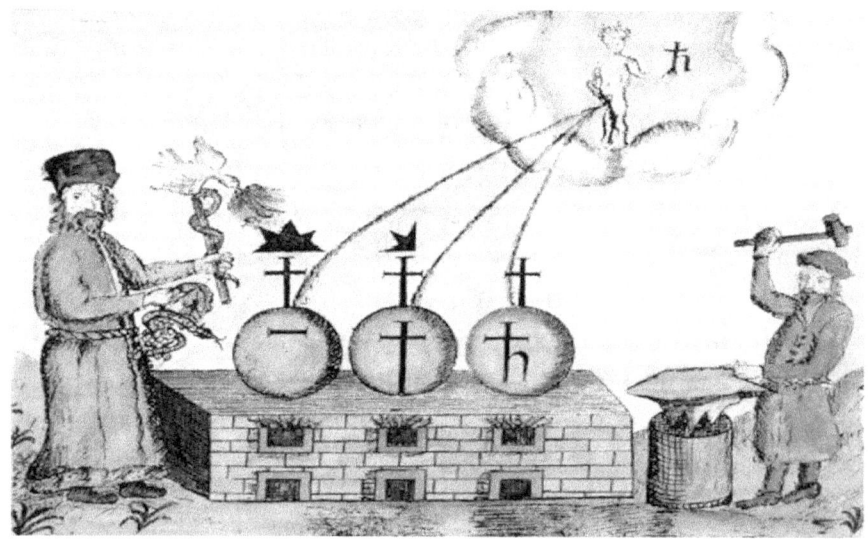

According to almost all good authors Saturn is the key to the alchemical process, the three jets of urine relate to the three regimes of the Great Work.

Alchemical Chronology

Before our era

– 20000 to – 10000
– Discovery of the first dyes in Lascaux rock paintings.

– 6000 to – 4000
– Discovery of copper (telllo) and antimony (Summer), gold being already known.

– 4000 to –3000
– Beginning of the Bronze Age. In Egypt we embalm the dead.

– 3000 to – 2 500
– Invention of gold purification techniques by coupellation in Babylonia.

– 1900 to –1800
– Using meteoritic iron.

– 1350 to –1250
– Supposed start of the Iron Age.

– 800 to – 700
– Manufacture of glass objects, enamels, use of antimonial salts, copper oxides, cobalt, tin.

– 350 to – 320
– Writing the manuscript of Thebes relating to the art of imitating gold and precious stones.
– Aristotle develops his physique of the four principles and "prima material".

– 200
– Archimedes invents the so-called double weighing method, the first non-destructive analysis to determine with precision the density of solids.

After J.-C

200 to 300
– Mariy the prophetess (Myriam) teaches alchemy at Zozime.
– Diocletian prohibits alchemy and burns all the books dealing with transmutation.

400 to 500
– Byzantines make up large compilations of works attributed to Marie, Zozime, Synesius, Olympodore.

500 to 600
– Alexandria Christians destroy the large library as well as most laboratories under the devastating authority of Archbishop Théophile.

650 to 700
– Morien, Hermite Romain, teaches alchemy to the Arab prince Chalid Ibn Jazid who will publish "*The Book of Cratès*".
– Calid writes "*The Book of Mysteries*".

720 to 800
– Jabir Ibn Hayyan (Geber) describes the salts of Arsenic and Mercury.

900 to 950
– Rhases experiences the minium, the litharge, the orpiment and develops metal sulfides.
– Avicennes in his "*Kitab as Sifa*" declares that if it is possible to imitate gold and silver, transmutation is not possible.
– Al Nadim publishes the Encyclopedia "*Kitab al Fihrist*".

1100
– Discovery of bismuth, sulfuric acid.
– Artéphuis, Arab author, signs "*Clavis maioris sapientiae*", and *Liber secretus*

1182
– Robert Castensis writes the "*Liberté composition alchemiae*".

1200 to 1250
– Michel Scott dedicates his "*Secretis*" to his master, the emperor Frédéric II.
– Albert Le Grand studies the action of nitric acid on metals, prepares potash salts.

1247
– Roger Bacon invents cannon powder.

1260
– Saint Thomas Aquinas takes up the Aristotelian philosophy of the *"primary material"*, but does not cultivate sacred art.

1310
– Arnault de Villeneuve operates a transmutation in front of the Roman Curia, he published *"Le Chemin des Chemins"*.
– Guillaume de Loris and Jehan de Mehung write *"Le Roman de la Rose"*, a masterpiece of alchemical poetry.
– Renowned alchemist others marked this era: Petrus Bonus (de Ferrare) (the good iron stone), Jehan de Ruspessica, the Ortholain, John Cremer.

1312
– Raymond Lulle and his namesake, write half a dozen small treaties. The *"Collarbone"* describes a particular process in which money and nitric acid play a very important role, it performs a transmutation in England.

1382
– Nicolas Flamel operates a transmutation of gold mercury on April 25, 1382. He writes in particular:
 "T*he book of hieroglyphic figures"*,
 "The Book of Wass",
 "The desired desire", "
 The philosophical summary"
 "The Bréviary".

1400 to 1500
– Nicolas Valois, Nicolas Groparmy and Vicot completed the great work in 1420.
– Jehan de La Fontaine publishes *"La Fontaine des amoureux de la science"*.
– Basile Valentin establishes the series of all antimony salts and develops medication against venereal diseases. He writes the
   "Twelve keys to philosophy",
   "The Treatise on Natural and Sternatory Things",
   "The triumphant chariot of antimony".
– Bernard Le Trévisan publishes two books:
   "Natural philosophy of metals"
   "The lost speech".
– Georges Ripley, English alchemist, Augustin religious, publishes the *"Book of Twelve Doors"*.

1500 to 1600
– Agricola wrote the first treaty of *"Re metallica"* mineralogy in 1530.
– Augurelle writes a famous poem *"La Chrysopée"*.

– Blaise de Vigenère writes an imposing treaty of salt and Solomon Trimosin publishes his "*Splendor Solis*".
– Paracelse is inspired by Basile Valentin to strengthen the notion of the third principle (salt) and occupies the first chair of chemistry at the University of Basel.
– Denis Zachaire writes "*The true natural philosophy of metals*"
– Latin edition of anonymous treaty "*The art of making gold, called chemistry*".
– Kelley and Jean Dee carry out public transmutations in Germany and England.

1600 to 1700

– Sethon, a Scottish follower, better known by the pseudonym of the cosmopolitan, is imprisoned by the Duke of Saxony, Christian II, for having refused to reveal the secret of the philosopher's stone. He escapes thanks to a Polish gentleman, Sendiviogus, and leaves a great classic in heritage: "*The new chemical light*".
– Stolcius makes the "V*iridarium Chymicum*" appear in Latin
– The President d'Espagnet writes "*The hermetic arcane of philosophers*",

1600 to1700

– Clovis Husteau de Nuysement writes his alchemical poems.
– Jean-Baptiste Van Helmont, a Flemish doctor and chemist discovers carbon dioxide, invents the thermometer and performs a transmutation using a powder given to him by an unknown.
– Pierre-Jean Favre writes his treaty "*Abrégé des secrets chimiques*".
– Henri Khunrath, alchemical doctor, composes "*The Amphitérish of Eternal Wisdom*"
– Valentin Andreae writes his "*Chymic Wedding*"
– Maier Michel publishes "*Fugitive Atalante*" and "*Intellectual songs on the resurrection of the Phoenix*"
– Huginus a Barma publishes "*The reign of Saturn changed in the golden century*".
– Philalethe, an American follower, writes the most popular treaty in the history of alchemy:"*Open entrance to the shut-palace of the king*", "*The Metamorphosis Of Metals*", "*The Celestial Ruby, "The Fountain Of Chemical Truth*".
– Isaac Newton studies alchemy and experiences in the stove.
– Limojon de Saint Didier comments on the treaty entitled "*The Old War of Knights*" in his "*Hermetic Triumph*".
– Helvétius made a transmutation in 1667, thanks to a powder given by a follower with the bizarre name of Dutch origin, *Elia Artista*.
– Barent Coenders Van Helpen published his treaty, "*The staircase of the wise*". In England, Robert Fludd systematizes Rosicrusian doctrines and is emulated in Germany: the shoemaker Jacob Boehme.

1700 to 1800

– Georges Stahl invents phlogistic theory.
– Bruno de Lansac (B.D.L.) comments on the wonderful songs of Marc-Antoine Crosslame "*The light coming out of oneself from darkness*".
– Dom Pernety writes his "*Mytho-hermetic dictionary*".
– 1732 anonymous treaty entitled "*Hermetic recreation*".

1800 to 1900

– Louis-Paul-François Cambriel publishes his "H*ermetic philosophy course in 19 lessons*".
– Cyliani achieves a cold transmutation of gold mercury and dedicates to posterity his "*Hermès Revealed*".
– Albert Poisson writes "T*heories and symbols of alchemists*".

1900 to 2,000

– Pierre Dujol, friend of Fulcanelli, published his "*Hypothyposis*", (commentary on the boards of the "*Mutus libe*r" of Jacob Sulat).
– Fulcanelli, last known follower, bequeaths two masterpieces of alchemical literature to his only disciple, E. Canseliet:
    "*The mystery of Cathedrals*"
    "*The Philosopher's Houses*"
– Claude d'Ygé, disciple of Canseliet, publishes his "*New assembly of chymic philosophers*".
– Eugène Canseliet, the most prolix alchemist author of the 20th century weakens an important heritage: with "*Two alchemical houses*", comments of the "*Twelve keys to philosophy*" of Basile Valentin, "*Alchemy*", Comments of the "*Mutus Liber*", writes "*Three former alchemy treaties*" and "*Alchemy explained on its classic texts*".
– Eugène Canseliet has also prefaced the works of many contemporary authors.
– René Alleau, author of "*Aspect of traditional alchemy*", directs a collection of unpublished alchemical works in the 1970s.
– Bernard Husson, outstanding hermetist, excellent multilingual, historian in addition, bequeathed us precious comments on several reissued works:
    "*Two alchemical treaties of the 19th century*",
    "*Anthology of alchemy*",
    "*Chymical Verger*" (translation from the work in Latin).
– Atorene, friend and disciple of Canseliet, excellent practitioner, publishes an undeniable practical scope study: "*The alchemical laboratory*".

Alchemy and Religions

We have just seen that alchemy was not of purely Egyptian origin, but more likely Babylonian. It is obvious that the Jewish contribution was decisive in the development of this science until the 5th century AD, a period from which Arabs and Christians resumed the torch.

Until the 8th century, alchemy was above all a metallurgical discipline which mixed fire arts and superstitions in close correlation with secular shamanic practices. The hermetic, spiritual or philosophical deviation which will then intervene is the fact of the continuous intrusion of the mystical and religious Christian currents and perhaps also partly Muslim (Sufis), from the beginning of the Middle Ages.

This mystical infiltration is especially noticeable in the speculative and pseudo-theoretical part of hermetic treaties, because it is it which makes all these works with a strong religious connotation incomprehensible, the theoretical pseudo part having nothing to do with the operating process.

This is why all modern researchers and historians, without carrying out the in-depth studies that this subject deserves or its patic experimentation, immediately put alchemy in the category of mystico-spiritual curiosities. They are thus removed any scientific and material scope from alchemy.

However, if some mystical alchemists wanted to spiritualize this art, mixing practice and religion, they did it, not to veil what was already obscure, but rather to claim belonging to a certain elite, to sects or to secret societies, like those of the Rosicrucians and Freemasons for example.

This observation is confirmed by Mr. Caron and S. Hutin, who wrote: "In the 14th and 15th century European alchemy tied to constitute itself in secret gnosis, claiming to give an airtight interpretation of the dogmas of Christianity.

In this gallery of mystical authors, we can store Mynsicht, Khunrath, Valentin Andreae, Fludd, Jacob Boehme, Maier, Grillot de Givry, illuminated authors or hermetists from the 20th and 21st centuries.

18th century chemistry laboratory.
Equipments of this laboratory all come from alchimic research.

These mystics have built a bridge between alchemy and the Christian religion to endorse a sectarian vision of hermeticism, like some of their precursors as Givry Grillot for example.

For these authors, alchemy is the realization of the evangelical message. Here's how they develop their concepts:

– the raw material is assimilated to the unmoved virgin which must give birth,

– calcination relates to the death of Christ and to the mystery of the cross (the cross appearing the crucible), and the iron, used for the reduction of ore, symbolizes the three nails of crucifixion, or the spear of Longin,

– the purification of matter is assimilated to the resurection of Christ,

– the ternary of alchemical principles (mercury, sulfure and salt) is compared to the Holy Trinity etc.

For the sect of the Brothers of the City of Hélios (FCH) created by the disciple of Fulcanelli, alchemy is a gift from God (*Donum Dei*) reserved for a deserving elite that fervent faith and work in the stove will reward.

Understand that Muslims, Jews or agnostics will simply be excluded from the Circle of followers. It is very unfortunate for non-Christian precursors who are the only inventors of this science and the only vectors of progress, in Europe and until the first millennium, at least.

Thus, certain communities will undergo insidious racism, intellectual discrimination. It is all the more ridiculous since the promoters of this racism have cheerfully hacked the know-how of the very people as they exclude from their spheres.

The truth, as Berthelot demonstrated, is that alchemy is gradually constituted from the foundations established undoubtedly in Mesopotamia in high antiquity, exported and developed by the Jews in the Mediterranean basin, then finally broadcast by the Arabs in the Middle Ages. So why this stupid religious "*hold up*"?

This grotesque intellectual hacking must be put an end. Alchemy is in no way a religion and does not belong to any particular religion.

It's an art, and like all arts, alchemy is universal. If imagiers, in the Middle Ages, immortalized hermetic doctrines on the walls, stained glass and statues of Ghotic cathedrals, it is only in order to secularize the operating techniques of the old chemistry and not to serve to serve Doubtedly kind of religious propaganda ends. In the Middle Ages, the great masters used Greek and Egyptian mythology as well as allegory and cabal, with a lot of spirit, and they showed a singular learning to veil their teaching. Thus, Dom Pernety explained at length in the Greek and Egyptian fables unveiled and in his mytho-hermetic dictionary the relations associating the process of the great work with the fables of ancient mythology. This idea had already been issued by the scientist Jacques Toll, around 1635 and this is also what the Bernard Husson scholar explains in "*Anthology of alchemy*".

The mythological encryption technique used by the ancient masters consisted in associating, with each of the materials of the great work, a character, a fabulous animal or a divinity. The fables involving the actors thus defined, were supposed to represent one or more of the operating phases of the alchemical process. The works of Hercules, the quest for the Golden Fleece with the Odyssey of Argonauts allegated all the operations required to obtain the philosopher's stone. In the alchemical bestière we will recognize the griffin, symbolizing the result of the first conjunction, the seven head hydra which appears the seven eagles or philosophical sublimations and the Phenix, emblem of the completed stone, reborn from its ashes. It will especially note, that it is not necessary to invoke characters from any religion, to veil the alchemical process.

Mention and encrypt the different stoves in the stove was not enough, it was also necessary to mark out the process of benchmarks easily identifiable by the disciples.

In order for the secrets to be lost, the former masters had the kindness of reporting the many clues that appear during the diets.

These are, for example, colors, the physical aspect of matter, a smell, a distinctive mark or sometimes a sound. Mythological encryption represented a powerful means of camouflage: to disclose so that the secrets are not lost, but to veil so that they do not fall into unworthy hands, that was one of the main concerns of followers.

The richness of the fables of antiquity and all possible combinations with allegory or hermetic cabal allowed the dissemination of secrets so as to transmit knowledge only to scholars and intellectuals which had the good keys, those of *the language of birds*, *the hermetic cabale* or *slang*, as Fulcanelli demonstrated.

But learning all these styles of cabal and esoteric language is not enough. In truth, the researcher must tirelessly pursue experimentation in the stove, in order to establish the link between alchemical symbolism and the metamorphoses of the compost.

Alchimy and Spagyria

Alchemy has often been confused, a secret discipline which deals with the transmutation of metals and the search for philosopher's stone, and spagyria, a very likely ancestor of our modern chemistry. Fulcanelli[4] suited, with reason, and developed this thesis, in "*The philosopher's houses*".

Spagyria is mainly concerned by practical and concrete developments in the laboratory and uses trivialized materials, while alchemy advocates the use of canonical materials of secret properties and involves unknown energies, conventionally called "*spirits*", in certain reactions . Spagyria is the branch that relates to the development of drugs from plant, organic or mineral extracts. It intervenes in industrial manufacturing processes (dyes, soap, sugar), in metallurgy (purification of metals and production of alloys), and the glass industry.

If many spagyrical transmutations have been reported in the past, it was limited, on specific substrates, prepared by specific processes, that is to say non-alchemical.

Another branch, according to Fulcanelli, was also dealing with transmutations, the archiemia or the architemospheria. One of the craftsmen of this branch would have been formed by a Tunisian scholar. It is Vincent de Paul, first a slave and then freed, who returned to Europe with an infallible recipe to make gold from silver in sheet subjected to the action of a mysterious powder. This process strangely resembles those developed by the American Emmens and the French Tiffere[5].

Learning alchemy must go through that of spagyria. No manipulation in the laboratory is possible without the perfect knowledge of the chemical properties of the materials implemented. This precaution remains valid today, because of the dangers of poisoning or explosion that the impetrant runs. In this regard, it would be good to consult the excellent work of Lemery[6].

Alchemists or Blowers?

The alchemical corpus has many incomplete, truncated, unfinished or altered works. It also includes an incredible number of books written by false alchemists, brilliant illusionists, in a word of the blowers who copied the style and expressions of the great masters. These works are a real danger for the uninitiated, because in addition to the useless loss of time and money for those who try to reproduce convoluted experiences, the risks of explosion or poisoning by toxic emanations for having literally applied false recipes are unfortunately common.
When we talk about ancient manuscripts or alchemical grimoires, it should not be believed that they constitute an instantaneous pos-

sibility of popularization. These texts are encrypted, coded, faked, they were only written for initiates.

Sometimes they include rebus or anagrams more or less easy to decipher or whose meaning has rarely been revealed by a charitable commentator. In other cases, it is a question of solving real puzzles whose meaning can only be revealed if you have acquired the perfect mastery of the magisterium in the laboratory.

One of the most extraordinary aspects of alchemical illustrations relates to the didactic character of symbolism, which is the imagined reflection of the incriminated doctrine point. René Alleau[4] had an admirable sentence: "*In alchemy, he says, everything that is symbolic is observable, as well as everything that observable is symbolic*". Indeed, to understand the alchemical symbols, you must have experienced yourself in the oven.

These revelations are in fact intended only for other initiates.

It's just frustrating. How to find your way in a faras of very pictorial, poetic and irrational words or expressions like *"The lascivious Venus, the embrace of Diane doves, the spirit of mercury, the philosopher's mercury, the philosophical mercury, the mercury of the wise, The mercury of the philosophe, the simple, double or triple mercury, the mercury of mercury, the philosophical gold, the pontic water, the dry water which does not wet their hands..."*?

As several branches of our old chemistry have cohabited over the centuries with alchemy, confusion voluntarily maintained in the terminologies used make it even more difficult to attempts to investigate.

That's not all, there's not just one alchemical process. There are several processes to lead to weight transmutations. These so-called "*individual processes* " - as opposed to the so-called "*universal alchemical process*" - are sometimes implemented. These "*individuals*" are mostly developed using materials specific to the work, but do not follow the entire alchemical process. They generally give more mediocre quantitative and qualitative results.

As for the purely alchemical processes, there are two ways with some variants - which also dispute certain followers by envy or by ignorance.

How to explain this mania to encrypt alchemical operations?

The distant precursors, Babylonians, Jews, Greeks, shamans recognized or raised to the rank of high priests and often protected by the emperors and kings of these eras, by political or economic strategy, invented religious ceremonials as well as jargons reserved for their caste, For each scientific field whose secrets they jealously retained. By caution, it was necessary at all costs to prevent any popularization assimilated to a sacrilege and to prevent secrets from falling into enemy hands. According to Mircea Eliade, these ceremonies include the choice of certain months to melt alloys, depending on the zodiac, as well as animal sacrifice and shamanic incantations. In any event, it seems certain that alchemists are the inventors of cryptography.

The development techniques as well as the airtight theories were chased as new processes were experienced or developed. Of course, failing any rational evaluation system, analogy was carried out by drawing parallels between the three kingdoms, animals, vegetable and mineral.

At the very beginning of the history of alchemy, there were probably many counterfeiters and charlatans who tried to make alloys imitating gold or silver and who falsified the titles of currencies with more or less happiness. There is a large number of manuscripts, grimoires or other works containing completely eccentric recipes, having absolutely nothing in common with alchemy. Thus, for example, this book entitled Le Petit et le Grand Albert, falsely attributed to Albert Le Grand remains a model of the genre.

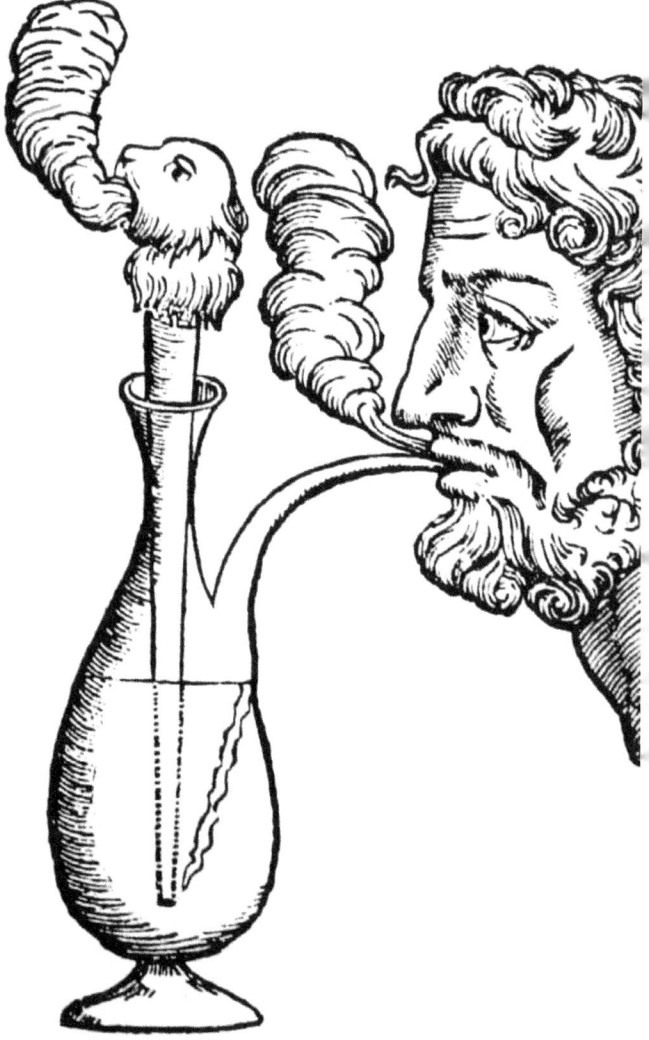

The false transmutations attributed to these blowers did not delay many people. In the Middle Ages, the touchstone quickly made it possible to determine with good precision the title of gold and the technique of double weighing was known for a long time. Besides, the followers were not embarrassed to denounce fraudsters and warn their readers against these practices.

One of the most common frauds that the followers denounced consisted in precipitating the copper of an aqueous solution by an iron blade, immediately appeared a cloud of golden precipitate having the appearance of gold. Nowadays such properties are used by conjurers in towers still illusing.

It should also be remembered that a large number of works are attributed, wrongly, to real alchemists, even if they are the work of renowned authors, and that others, no less numerous, only deal of specific processes

*
* *

---

1. FULCANELLI, Le Mystère des Cathédrales, (*The Mystery of the Cathedrals*), third edition, Paris Jean-Jacques Pauvert, 1964.
2. Eugène CANSELIET, Alchimie, études diverses de symbolisme hermétique et de pratique philosophale. (*Alchemy, various studies of hermetic symbolism and philosophical practice*), Paris, J.-J. Pauvert 1964.
3. Claude d'YGÉ DE LABLATINIÈRE, Nouvelle Assemblée des Philosophes Chymiques (*New Assembly of Chemical Philosophers*), Paris, Dervy Livres 1954
4. FULCANELLI, Les Demeures philosophales (*The Philosophers' Mansions*), third edition, Paris Jean-Jacques Pauvert. 2 vol
5. VEZE, Marcus, L'Or Alchimique (*Alchemical Gold*), Lyon, Éd. du Cosmogone, 2001
6. LEMERY, Nicolas, Cours de Chymie (*Chemistry Class*), Paris, Houry, 1757

Chronology of Some Transmutations[1]

1370 BC.
According to a cuneiform tablet, almost all of the gold provided by the pharaoh Amenopis IV to the King of Babylon did not resist the test of fire. It's the oldest known fraud Hsitoire.

1310
Arnauld de Villeneuve performs a transmutation in front of the Roman Curia in Avignon.

1400 - 1500
Nicolas Flamel succeeds in a transmutation on an Easter Monday.

Ripley, English Augustine religious, finances the military resistance of the Knights of Rhodes with alchemical gold.

1500 - 1600
Vincent Depaul says he learned archimal secrets during his captivity in Tunisia.

Auguste de Saxe fills his chests with alchemical gold produced by a certain Beuther.

Agricola testifies to a transmutation in an Italian convent in 1630.

Becher makes a silver medal transmuted from lead.

Richthausen, alias Le Baron du Chaos, has a dye, part of which transmmed 16,470 times its weight of golden mercury.

The Count of Rusz, general engineer of the Mines of Styri and Carinthia, publicly operates a transmutation in 1648, using the dyeing of Richthausen.

Johann-Philippe de Scönborn, voter of Mainz, transmutes two pounds of Mercury in gold in front of Baltazar de Monconys, in 1664.

Brechtelt, goldsmith of The Hague, reports his testimony on the transmutation made by Helvetius in 1666, thanks to the powder given by the Dutch follower Elias Arta.

Spinoza endorses the authenticity of the transmutation operated by Helvetius.

Monte Snyders, hermetic philosopher, operates successful transmutations and cures in Vienna in 1666.

Georges des Closets performs a transmutation of 8 pounds of gold in gold in the presence of Jean Vauquelin des Yveteaux.

Hauton, a Norman doctor witnessed a rejuvenation due to the ingestion of an alchemical remedy.

The Baron de Waghnerck made in 1680 a golden lead transmutation in the goldsmith Charles Le Blon in Frankfort, the powerful power of his powder was of one in 5,250.

Khon, dean of the Ulm medical college, witnessed a public transmutation on June 12, 1695.

Desnoyers alias Plantier, owner of a projection powder, acquires immense properties around Paris.

1700 - 1800
Von Paykel transmutes lead into gold in 1706.

Kundmann, doctor of medicine in Breslau, attends a gold transmutation of an amalgam of lead and mercury in The Hague, in 1708, the alchemical gold aloine was experienced by the chemist E. Stahl.

Kington, an English alchemist, operating under the pseudonym of Pierre de Leyde, transmutes 177 pounds of Mercury in gold which was sold by a certain bournet to the bank of Amsterdam, in 1712.

Ernest-Louis de Hesse-Darmstadt had a hundred silver thalers hit in 1717 and several hundred gold ducats, obtained by lead transmutation thanks to two dyes sent by an anonymous sender.

Stahl analyzes a fragment of alchemical gold in 1708 and confirms the presence of a surplus of dyeing within the metal.

Geelhausen, professor of medicine at the University of Prague, carried out a public transmutation of silver lead in 1728.

Cyliani, a French follower, declares that he has carried out a "cold" transmutation, from golden mercury.

Reussing, a pharmacy trainer, made a transmutation in 1752 in Halle.

Christian Eisenbeg of Saxe-Gotha has a florin hit on archimical gold.

De Bournet relates the transmutation of 177 pounds of Mercury in gold made before his eyes in 1712 in Amsterdam.

1800 - 1900

Emmens, an American chemist, produced gold from Mexican silver, the gold thus produced is sold.

Tiffereau transforms money in Mexico, from certain silver ores.

1922

Canseliet performs a golden lead transmutation in front of two witnesses, Julien Champatre, painter and friend of Fulcanelli, and Gaston Sauvage, chemist at Poulenc, with a sample of projection powder developed by Fulcanelli.

---

1. Bernard HUSSON, Transmutations Alchimiques (*Alchemical transmutations*).

Paris. 1974.

Esotericism and Initiation in the 20 - 21st Century

Your first contact with alchemy came through the study of the Fulcanelli1. Unlike many esoteric authors, Fulcanelli[1] is a fascinating writer to read, his style is of unparalleled purity and he manipulates the word like the greatest classical authors.

Fulcanelli is an amazing author who masters languages, history, science and technology. His extraordinary skill will force your imagination to the point of considering the most improbable solutions, and will lead you into the worst dead ends.

To better understand the Adept of the 20th century, it is necessary to supplement his teaching by reading the works of his only disciple, Eugène Canseliet[2]. In a much less direct manner, with a rather emphatic style borrowed from previous centuries, he set out to decipher all of his master's work.

Once the study of Fulcanelli and Canseliet has been undertaken, when one thinks that one has finally acquired a satisfactory knowledge of Alchemical Philosophy, the next step consists of filling in the gaps intentionally left by the great Master and his disciple.

Fulcanelli, in order to excite the reader's curiosity and push him to discover the secrets hidden in other authors, tirelessly cites, to support his statements, numerous passages borrowed from the alchemical corpus. Instinctively, we are tempted to consult all the works indicated to complete voluntarily truncated information. What a desappointment! A beginner will find absolutely nothing that is not worthy of interest, at most vague clues scattered here and there, enough to make us stamp our feet in dismay.

The height of encryption for the Adept is to direct you towards other authors more envious than him.

Moreover, you have now understood that all Philosophers write under the seal of secrecy, that all their books are closed, but that one book can sometimes open another. When you accomplish these literary escapades which will help to complete your knowledge, you will become accustomed to the language of the ancients. This is what the grand master found best for your training. You will then learn to reason like an alchemist, speaking the Hermetic language. This is the only way to progress in this discipline from another era due to lack of an initiator.

In order to save precious time, it is in the beginner's interest to obtain works published by good contemporary authors, such as Claude d'Ygé[3], Bernard Husson[4] or Atorène[5]. He will quickly realize that the deeper we go into the past, the more difficult the assimilation of texts becomes. An important detail: knowledge, even basic, of dead languages such as Latin or ancient Greek, is a major asset in this art. Over the years, the amateur will accumulate a significant collection of works. Fortunately, there are many charitable publishers who regularly bring back from the past masterpieces and classic treatises of the Chymistry of the Ancients. The word charitable, whose use in alchemy has become commonplace, is entirely appropriate. Nowadays you can't get rich by publishing books with fewer than a thousand copies. Let us salute in passing the long chain of those who contribute to perpetuating the Ancient Tradition: authors, translators, commentators, publishers and booksellers.

Once the study of contemporary authors has been completed and which constitutes only an appetizer, we must tackle the main course: the alchemical corpus of past centuries.

How can we define the criteria of choice in a field where literary critics are conspicuous by their absence? Nothing should be ruled out a priori. You have to read everything and consider that this is the price to pay to complete this learning.

What is important to know is that we are very often confused by reading completely hermetic texts and upon examination of which we would be tempted to hastily conclude that they have no connection with alchemy. . The authors of this type of work take pleasure in stylistic effects in order to amaze and mislead the reader. This literature resists all examinations because it is so obscure.

Some books were written by so-called initiates or mystics who frequent strange sects. This is the case of the two books by Grillot de Givry[5,] one entitled *"Le Grand Œuvre"* and the other *"Lourdes"*. Their study does not allow us to draw any positive lessons. The first book looks more like a collection of prayers than a treatise on alchemy, the second is entirely dedicated to Marian worship. However, we must be honest, the *"Museum of Sorcerers, Mages and Alchemists"* by this same author is a work worthy of interest.

Also be wary of authors who, pretending to write under the cloak of philosopher, propagate fallacious theories. Through subtle suggestions, they will make you doubt your knowledge and lose all the knowledge you have painfully acquired. AvoidValois, Nicolas - Urbiger

Here are other titles by anonymous authors, as interesting as they are rare:
— Hermetic recreations, followed by Scholies
— The Light emerging by itself from the darkness,
— The Psalter of Hermophilus sent to Philalethes,
— The Philosophers' Rosary,
— Written science of all hermetic art.

To avoid unnecessary waste of time, you will benefit from consulting good contemporary authors.

I will first of all cite my true initiator, Bernard Husson[8], accomplished hermeticist, great initiate, excellent linguist, honest and humble scholar. Husson has unearthed real treasures of alchemical literature. In his *"Two Alchemical Treatises of the 19th Century"* he restored the entirety of the *"Hermetic*

Recreations", an unpublished alchemical manuscript of major importance for the hermeticist, the certain inspiration for the work entitled *"Hermes Unveiled"* attributed to Cyliani. He was thus able to put a name under the initials B.D.L., the erudite commentator on *"The light emerging by itself from darkness"* – the Adept Bruno de Lansac – and he wrote an excellent *"Anthology of Alchemy"*, masterfully commented on the *"Viridarium Chimicum"* as well as the work entitled *"Discourse of an Uncertain Author on the Stone"*. We also owe him an extremely interesting and perfectly documented study on the history of transmutations. Bernard Husson, who died at the end of the previous century, had a perfect command of Greek, Latin, German and English. His various talents allowed him to translate and comment on numerous works, notably for the collection directed by René Alleau, the *"Biblioteca Hermetica"*.

Bernard Husson was stripped of the authorship of his discoveries by some contemporary authors. He had dared to challenge Fulcanelli, in a study relating to the statue of Saint-Marcel slaying the dragon during the reissue of Cambriel's *"Course of Hermetic Philosophy in Nineteen Lessons"*. These unscrupulous *"colleagues"* have often published entire passages borrowed from Récréations Hermétiques, revealed the character who was hiding under the initials B.D.L (Bruno de Lanzac), without naming the author of the work from which these extracts were taken, namely Bernard Husson.

But despite this disgrace, Eugène Canseliet reconciled with the eminent Hermeticist and he was one of the few to mention him; In his commentaries on Mutus Liber, then in his Alchemy explained on his classical texts:

"To our knowledge," writes Canseliet, *"it was Bernard Husson who brought it to light in his first work. It is therefore for the second time that our friend published in full, at the end of his Anthology of Alchemy, this probably autograph booklet which its completely unknown author titled, not without the cabalistic*

humor of the artists, by the fire: Hermetic Recreations."

The second contemporary author worthy of interest is only known to me by his pseudonym Atorène. He was a close friend of Eugène Canseliet, with whom he maintained a regular correspondence. Atorène[9] only wrote one book entitled "*Alchemical Laboratory*". It is the work of an authentic initiate, rich in practical details that few Adepts have dared to reveal until now. This book is very pleasant to read and contains many illustrations. It is referenced in the bibliographic index at the end of the book "*Alchemy*"[10], published by Dervy. On this occasion the author of this index congratulates Atorène for "*his profound spagyric knowledge*". On this subject, I questioned the author of the very substantial and original bibliography which appears in this work, Mr. Richard Caron, in order to clarify the reason for the epithet "*spagyrics*", used in the circumstances. Mr. Caron, with extreme kindness, let me know that it was necessary to take this word in its medieval meaning, as Basil Valentin or Paracelsus liked to use it, which amounts to complimenting, very elegantly, Atorène, for the quality of his teaching. There should therefore be no mistaking Atorene's real alchemical skills.

Finally, I would also recommend the study of two other contemporary authors that I like, their books contain excellent deductions. These are Jacques Sadoul[11] and Fabrice Bardeau[12].

---

1. FULCANELLI,
    - Le Mystère des Cathédrales (*The Mystery of the Cathedrals*), third edition, Paris J.-J. Pauvert, 1964
    - Les Demeures philosophales (*The Philosophers' Mansions*), third edition, Paris J.-J. Pauvert, 1965 2 vol.
2. Eugène CANSELIET,
    - Deux logis alchimiques (*Two alchemical dwellings*), Paris Jean Schemit,, 1945
    - Alchimie (*Alchemy*), Paris J.-J. Pauvert, 1978,
    - L'Alchimie et son livre muet (*Alchemy and its mute book*), Paris J.-J. Pauvert, 1967,

- L'Alchimie expliquée sur ses textes classiques (*Alchemy explained from its classic texts*), Paris J.-J. Pauvert, 1972,
- Trois anciens traités d'alchimie (*Three ancient alchemy treatises*), Paris J.-J. Pauvert, 1975.

3. Claude d'YGÉ de LABLATINIÈRE, Nouvelle Assemblée des Philosophes Chymiques *(New Assembly of Chemical Philosophers)*, Paris, Dervy Livres 1954.

4. Bernard HUSSON,
- Deux Anciens Traités d'alchimie du XIXe (*Two old Treatises on Alchemy from the 19th Century*)   Paris, Omnium Littéraire, 1964.

Contains the following treatises:
- CAMBRIEL, Cours de Philosophie Hermétique en 19 leçons (*Hermetic Philosophy Course -19 lessons*), 1843,
- CYLIANI, Hermès dévoilé (*Hermes unveiled*), 1932,
- Anonymous, Récréations Hermétiques (*Hermetic Recreations from the 18th Century*)
- Anthologie de l'Alchimie (*Anthology of Alchemy*), Paris Pierre Belfond 1978,
- Viridarium Chimicum, Paris Librairie de Médicis, 1975,
- Transmutations Alchimiques (*Alchemical Transmutations*), Paris, éditions J'ai Lu, 1974,
- Discours d'auteur incertain sur la pierre des philosophes (*Uncertain author's speech on the philosophers' stone*), Dervy 1996,

5. GRILLOT de GIVRY, Émile-Jules
- Le Grand Œuvre, (*The Great Work*) Paris, Chacornac, 1960,
- Lourdes, ville initiatique (*Lourdes, initiatory city*), Paris, Chacornac, 1930,
- Le musée des sorciers, mages et alchimiques (*The Museum of Sorcerers, Mages and Alchemists*), Paris, Lib. de France, 1929.

6. Dr Marc HAVEN, The Mutus Liber, Paris, Chacornac, 1914.

7. Eugène CANSELIET, L'Alchimie et son livre muet (*Alchemy and its mute book*), opus cit.

8. Bernard HUSSON, opus cit.

9. ATORÈNE, Le Laboratoire alchimique (*The Alchemical Laboratory*), Paris, Trédaniel-La Maisnie, 1982.

10. Richard CARON, Bibliographie des ouvrages consacrés à l'alchimie publiés en langue française de 1900 à 1995 (*Bibliography of works devoted to alchemy published in French from 1900 to 1995*). In alchimie.

11. Jacques SADOUL, Le trésor des alchimistes (*The treasure of the alchemists*).Paris, Denoël 1970.

12. Fabrice BARDEAU, Les clefs secrètes de la chimie des anciens (*The secret keys to chemistry of the ancients*).Paris, Robert Laffont 1975.

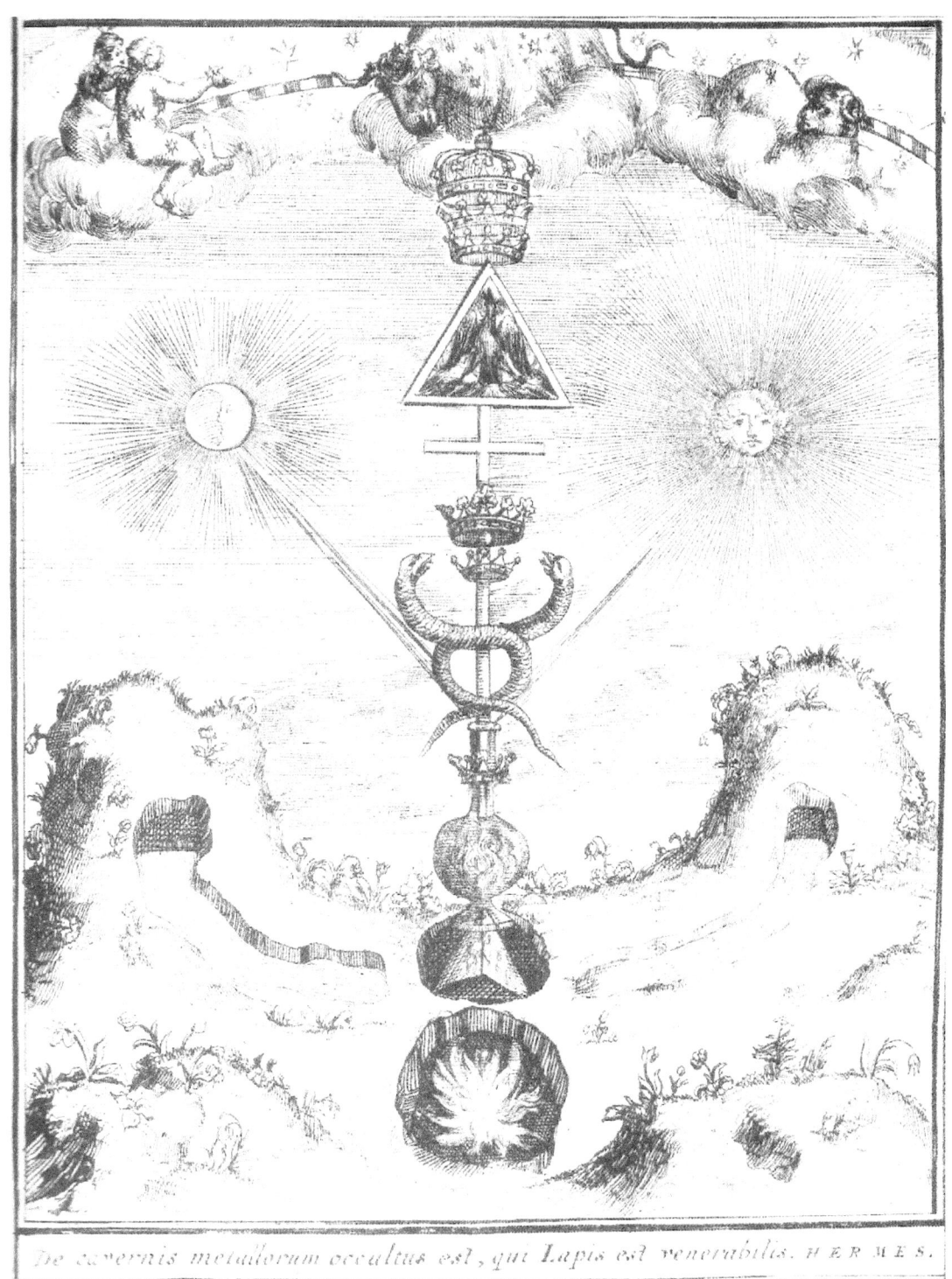

The venerable stone is hidden in the caves of metals (Hermes)

Birth and evolution of a theory

Alchemy was born in the hell of the metallurgists' forges, the goldsmiths' ovens and in the furnaces of the glassmakers and potters of ancient times.

It was by seeing metals in fusion, by melting tin, copper, silver, gold or antimony, by applying empirical refining processes, that the first beliefs naturally emerged. Mineral or metallic matter seemed to be born, live and die, just like the animal or plant kingdom in this lower world, but at an imperceptible rhythm, so slow did the process seem. Minerals, for these precursors, had to develop following a mysterious cycle, in the natural laboratories of their geological deposits, and under the influences of the astrological planets presiding over their destiny.

According to the doctrines that were gradually put in place, we came to imagine that metals, living bodies, were all destined to evolve towards perfection, the incorruptibility of gold being the final stage of mineral evolution.

Researchers, practitioners or scholars from these ancient times focused on the study of the phenomena caused by the materials used, during the operations of refining metals, manufacturing glass, acids, bases or lime. They set out to explain the transformations undergone by matter in order, perhaps, to rationalize production and develop new techniques.

It was necessary, first of all, to constitute an analog database, to organize the materials used by category, and likewise, for the reactions observed, to classify them into groups or families of easily identifiable reactions.

This is how the first definitions directly associated with the main states of matter were born, those of the Elements:
- the Earth, attribute of everything that is solid,
- Water, defining any liquid state,
- Air, the gaseous state,
- Fire, considered as an element, prefiguring the notion of energy.

The four elements were associated with physical properties that the senses could appreciate:
- the hot,
- the cold,
- the dry,
- the humid.

These definitions were not sufficient to describe all the phenomena resulting from the modifications generated at the level of matter by chemical reactions. The ancients therefore added the notion of "*Principles*":
- Mercury,
- sulfure,
- salt.

Finally, unable to explain certain metamorphoses, due to lack of appropriate analytical means, the alchemists then established the concept of Spirit to explain the dynamism of life and everything that is capable of causing transformations.

The four Elements, the four Properties, the three Principles and the Spirits, nothing was missing to describe with a certain reliability all the physical modifications that they could impress on matter.

Moreover, current science recognizes three states of matter, plus a pseudo "*energetic*" state with the relativistic mass-energy equivalence ($E=mc^2$). On the other hand, it gets a little complicated at the level of principles because, in fact, we cannot objectively establish equivalence between our acids, our bases and our salts – resulting from combinations of the first two – and Mercury, Sulfure and Salt of the old Masters, except the "Spirits" which should not be confused with the acids or bases of the spagyrists.

Some theoretical authors or hyperchemists have not hesitated to make clever combinations between earth-water, fire-air, hot-cold-or dry-humid to develop a purely symbolic alchemy. They even created childish assemblages of geometric figures and mixed astrology and magic. These combinations, sometimes called cabalistic, as subtle as they are, do not bring any clarification to the operations of the Great Work, they are incompatible with the great principles that we encounter in most hermetic axioms and aphorisms and do not withstand a deep analyze.

By studying the ancient texts in a complete and repeated manner, we can deduce, as the great learned chemist Chevreul1 did in his time, that the speculative part of alchemy has no connection with practice in the furnace and does not correspond not to what we would call today the theory of a science. It was actually not by combining our Mercury with our Sulfure (both philosophical) that the alchemists obtained mercury sulphide, cinnabar!

Chevreul[2] adds that this practice, while evoking the making of the philosopher's stone:*"consisted of giving life to an inorganic matter by operating the conjunction with a soul, through the intermediary of a spirit, an average substance participating in both matter and the soul"*.

The famous chemist thus, in very few words, masterfully exposed the nature and purpose of all Hermetic Philosophy. He is right to say that all the theories developed in alchemical writings, from Antiquity to the French Revolution, are perfectly out of step with practice with which they have absolutely no connection.

Sometimes these theories are touching in their naivety. Often so-called philosophical concepts are borrowed from Platonic or Aristotelian schools, as a pure exercise of style or to serve as intellectual support. This is what Arab authors did until the tenth century. and even certain European authors, such as from Spain[3] for example.

Alchemists from the Middle Ages until the 19th century believed in the existence of "*Universal Spirit*", a sort of energetic or magnetic agent, invisible, attached to life, having the power to condense and accumulate in certain bodies specially prepared for the Great Work. It was also a secret fire which, excited by external heat, would have the power to cause modifications in the structure of matter, to the point of giving it an incredibly high density. This secret fire, enjoying catalytic properties unknown to current science, would only operate perceptibly under certain conditions and at certain times of the year, according to some authors.

Ultimately, as some believed to deduce from the Emerald Table, the operative alchemy consisted of bringing the Upper Universe and the Lower World into perfect harmony by capturing and condensing a certain aerial spirituality, a spirit to use the correct expression , in a substrate capable of receiving it while retaining its undifferentiated nature.

The condensation of this universal agent causes, on the appropriate support, all the metamorphoses required to elevate the alchemical embryo to the level of the Philosopher's Stone.

All Hermetic Philosophy is found in these two verses taken from the Emerald Tablet:

"What is above is like what is below, just as what is below is like what is above."

At an indeterminate time, operational practice finds itself with two parallel processes leading to the same result: the Philosopher's Stone.

These two processes, the dry process and the wet process, themselves include numerous variations and sub-processes, but involve the same basic materials. For some authors, there is even a so-called *"brief"* third way.

We saw above that to veil the operational process, relatively simple for those who have a few keys and the right principles, most alchemists transmitted the main principles of the process in the form of puzzles, for example by inverting the order of operations, the names of materials. Even stronger, in order to further obscure the operating process, the dry method was very often described under the modus operandi of the wet method. Thus, the misled reader understands that it is necessary, at a certain stage, to use a liquid element, whereas we are talking about *"water that does not wet the hands"*, therefore a solid material. So much confusion!

After some research and a little common sense, we see that the entire process has been divided into three main phases called diets with certain similarities between them.

This encouraged the smartest authors to make deliberate mixtures between the three regimes and the two paths, and particularly Eyrénée Philalèthes[4] in his Introitus.

Alongside alchemy, it must be remembered that two other chemical disciplines appeared in the Middle Ages:

– archymy, chemical-metallurgical technique involving reactions caused by *"spirits"* on alloys of gold or silver with base metals such as copper, mercury or lead,

– spagyrics, a discipline in which Basil Valentin excelled, which was then taken up and developed by Parascelsus. Its main purpose was to make medicines.

This was also the pretext for some Adepts to mix spagyric manipulations with their alchemical treatise.

Indeed, there are some similarities between alchemy, spagyrics and archchemy, and, according to Bernard Husson, these three disciplines undoubtedly all appeal, although in different ways, to the *"universal agent"*.

Rabelais named this agent Entéléchie, adding very gentilely: *"shit whoever names it"* and there should be no mistake, it is the true pivot of Art.

Basil Valentin[5], speaking of the nature of the Stone in his Twelve Keys, assures that in nothing lies everything, but it is he the spirit or the universal agent, and it is also he, this famous light of nature to which Hermeticists of the past often allude to this. You will understand how easy it is to mislead the reader with this type of sequence.

This is what the many followers who preceded us had fun with.

Bernard Husson[6] will specify about this mysterious agent:

"While alchemy envisaged the accumulation of an agent on a temporary support but capable of effecting this condensation, spagyrists

were busy preparing remedies sometimes involving this universal agent, with more or less success, in compositions of extracts of plant or animal substances, just as the archimists involved this agent whose capture was immediately accompanied by a correlative transmutation, sometimes by an increase in weight of the precious metal initially present".

Here are the main points according to the alchemical axioms.

To alchemically obtain this condensation of the universal agent, it is essential to use only materials suitable for carrying out the Great Work.

In theory, these materials are found in all places, at all times, but if we examine the possibilities of nature, according to the fundamental axiom of alchemy, only ores and metals can participate in this great ontogenesis. We will therefore exclude everything that is animal, vegetable and organic.

In addition, we must reject everything that leaks or burns in fire as well as noble metals, because the latter have completed their cycle of transformation, and, in their state of perfection, they cannot communicate any virtue. Common mercury is, for its part, too young, and the specific agent that we are looking for is not found there, nature has never introduced it there.

The alchemical tradition also specifies that we must seek perfection in imperfect things, in base matter. In this nothing of Basil Valentin, or in this old dragon which infects everything with its venom, destroys imperfect metals and then converts them into more than perfect medicine.

In alchemical mythology, the Dragon is the symbol of the subject who dispenses Mercury, it is the black scaly matter that the artist must select at the beginning of the work and on which he must first work. This last sentence of the work, completely innocuous, conceals a big secret. This Dragon will then be defeated by the knight armed with the lance and the shield, according to Fulcanelli, in order to generate a little later the dolphin, or the remore, the only fish caught in the sea of the wise and which the artist must know how to season to make it reborn, like the Phoenix, from its ashes.

This way of expressing oneself is found in most classic works. Sometimes the alchemical work is presented in the form of a drama that matter must undergo. These are in no way fantasies born in the brains of men since the dawn of time as proclaimed by certain psychoanalysts who have no competence to criticize a chemical discipline, of which they are ignorant of the basic rules.

This drama was staged by Dante, who uses a subtle cabalistic style of encryption to veil the secret process. The artist plays the role of savior, coming to deliver the metallic soul to make it climb step by step the ladder of redemption, by successive sublimations, until the state of perfection which symbolizes the Philosopher's Stone. The secret process will then allow the rehabilitation of base or leprous metals which await resurrection behind the scenes, meaning their transmutation into gold.

In this animist approach, we subject matter to real torture in the crucible. We must kill the living and resurrect the dead to accomplish this great design. We will see later that another great manufacturing secret is hidden behind this axiom.

In alchemy, no sentence is devoid of practical meaning.

But it seems that there is no limit to exaggeration.

In other cases, we will assimilate the stone squarely to Christ, *"Our Stone must be enclosed in a container like Christ in the tomb"* (Arnault de Villeneuve), and it is unfortunately this vision borrowed from Catholicism that some Christian alchemists will convey for several centuries to such an extent that several authors seem to want to make alchemy a powerful aid to Christian proselytism.

The association of the alchemical process with the Christian Mysteries was perhaps originally intended to support esoteric research in the Middle Ages and to spare lovers of the Art the stakes of the Inquisition. This was no longer the case from the Renaissance onwards.

Today, fortunately, we no longer burn sorcerers, but bad habits have been adopted and there is much to be done for this discipline to become again what it would never have ceased to be: an open secular discipline. to all honest researchers.

Finally, to go from one extreme to the other, some authors have compared alchemical mercury to the Babylonian prostitute of the

Great Work. In alchemy, we move indifferently from the vulgar to the sacred or vice versa, without shame.

---

1. CHEVREUL, Journal des Savants (*Journal of Scholars*), Juin 1851.
2. CHEVREUL, Journal des Savants (*Journal of Scholars*), Décembre 1851.
3. D'ESPAGNET Jean, L'Œuvre secret de la philosophie d'Hermès (*The Secret Work of the Philosophy of Hermes*). Paris, E.P. Denoël, 1972.
4. PHILALÉTHE, Éyrinée, Entrée ouverte au Palais fermé du Roi (*Open entrance to the shut-palace of the king*), Paris, S.G.P.P. Denoël, 1970.
5. Basile VALENTIN, The twelve keys of philosophy, Op. cit.
6. HUSSON Bernard, Viridarium Chimicum, Op. cit.

Putrefaction - Fifth Key - Basile Velentin

Alchemical Glossary

– *Subject of the wise, Subject of Art:* one of the raw materials of alchemy in its raw state.
– *Magnesia:* another name for the subject of the wise.
– *Spirit of magnesia:* name of the hermetic solvent.
– *Raw material:* natural ore extracted from the mine.
– *Raw material*: pre-treated ore, enriched by liquation.
– *Common mercury* or simple mercury, or second mercury: regulates
– *Mercury :* female principle, first principle
– Mercury of the Sages or double mercury, philosophical mercury: material obtained at the end of sublimations, by the solution of philosophical gold.
– *Our Mercury:* flowing mercury (Hg) or hydrargyrum.
– *Mercury Triple* or Trimegistus: mercury used for the three works.
– *First Adam:* iron, knot of gold, sulfure, male element, extracted from the ferric slag of the first work.
– *Rebis, Rémore or Rémora*, Fish, Delphin: mercury resulting from the sublimations of the second work, having to undergo the final coction.
– *Doves of Diana:* mercury from the second work.
– *Sulfure:* male principle, second principle.
– *Salt:* third principle.
– *Saltpetre*, nitre salt: one of the saline adjuvants.
– *Harmoniac salt:* double salt, an ana mixture of niter and tartar, not to be confused with amonium chloride.
– *Tartar:* second adjuvant, confusion often maintained between tartar and potassium bitartrate.
– *Philosophical vitriol:* salt from the calcinations of the first work.

Some definitions to remember

— *Adept* designates the alchemist who has achieved the development of the Stone. Today this term designates a fan or enthusiast.
— *Artist*, or lover of Art, words which used to designate the alchemist.
— *Archchemy and spagyria* were cultivated by hyperchemists in the Middle Ages.
— *Athanor:* from the Hebrew Tanour, designates the secret oven of alchemists.
— *Caput Mortuum*, (dead man's or Moor's head), dark residue obtained after an operation of calcination in a crucible in the dry process, or distillation, in the wet process.
— *Compost:* designates the mixture ready to undergo cooking.
— *Compostela* refers to the star compost of the first work.
— The word *Philosopher* designates the scholar or alchemist of the Middle Ages (friend of science). Hermetic Philosophy formerly referred to alchemy, as did Chemical Philosophy.
— The *matra* is a glass flask used in the wet process.
— The *Blowers* constituted the arena of researchers who applied false recipes without taking into account alchemical axioms.
— *Reincrudation* is the term used in the past to designate the return of matter to a previous state. For Canseliet, reincrudation simply designated the fine grinding of matter.
— *Sorbonne:* glazed and ventilated niche in which chemists carry out operations likely to form toxic gases.

*
* *

Manufacturing Secrets

The two processes for producing the Philosopher's Stone (dry and wet) begin with the same operations, according to Fulcanelli.

Before starting the experimental phase, here is what you need to know:
    1 - the basic materials,
    2 - the quantities and proportions of the products used,
    3 - the necessary laboratory equipment and instruments,
    4 - the nature of the operations implemented at each stage of the process,
    5 - the properties of intermediate products and by-products,
    6 - the rules to follow at each stage,
    7 - the times required for each phase,
    8 - the quantities and proportions of intermediate products and by-products,
    9 - all phenomena observable during the process,
10 - the means of control,
11 - the rules of use for projection.

    Other authors would have added:
12 - knowledge of the right moment for each operation.

*
* *

Warning

As is customary, to support and reinforce the deductions presented throughout the chapters that follow, I will reproduce below numerous extracts taken from the best and most reliable authors.

These extracts alone, grouped into sections, already allow a first approach to the part accessible to any art lover. I could certainly have shortened this study, going straight to the point and avoiding certain repetitions, but this would have the consequence of removing all credibility from the revelations which follow, outside the bibliographic context.

In order to proceed with order and method, I will try to define, first of all, the basic ingredients in their raw state, the preliminary work called *"outside the work"*.

The manufacturing processes, strictly speaking, will come next.

By the admission of all the great Masters, the complete process of developing the Philosopher's Stone has never been revealed. Likewise, the preliminary preparations called "outside the work" were never divulged before the hermeticist Eugène Canseliet amply mentioned all of these operational details in his commentaries on the *"Twelve Keys"* of Basile Valentin, of the *"Mutus Liber"*, then finally in "Alchemy explained on its classic texts". If he had not done so, there is no doubt that these secrets would have been lost forever.

The main stumbling block lies in all kinds of confusions maintained by the best authors.

In alchemy everything is interchangeable, according to Bernard Husson's own expression. The Mercury of one regime could be the Sulfure of the other, Mercury could also be assimilated to Salt, which here becomes a principle and not the necessary mediator for the three works. Likewise, Salt will be taken for one of the fires, knowing well that there are other fires etc.

Bernard Husson has pointed out, on various occasions, the difficulty of resolving the mysteries of the Great Work by also demonstrating that the protagonists of the work change costume and name from one regime to another.

This is how, commenting on the first key of Basile Valentin (in "*Le Jardin Chymique*")

Bernard Husson points out this trap while offering a clue of inestimable value for the beginner:

"This is why, if four bodies, chemically defined, in the current sense of the term, each playing the role of one of the traditional cosmic elements, are actually required for the start of the operation mentioned in this first key, it is not necessary not believe that each corresponds unequivocally to one of the four figures on the board. It symbolizes in fact two similar and successive operations, the superposition or shift of which (in analytical decryption) represents the essential operative secret of all alchemy. The exposition process, deliberately sybilline, is exactly the same here as in the Introitus of the Philalethes. The warning that we are probably the first to formulate so explicitly, for researchers already advanced in the study of alchemical texts, also applies to the character on the crutch, carrying a scythe and carrying a cup of refining, both Saturn and Vulcan, as for the wolf leaping over the Hessian crucible surrounded by flames. Without excluding its well-known spagyric meaning (where the symbol, losing its multiplicity, is reduced to a univocal allegory, and no longer interprets in a perfectly satisfactory way the text which explains it in a deliberately contradictory way), the gray wolf is also here the green wolf, whose festival was still celebrated, until around 1830, near Jumièges, on the eve of Saint-Jean."

*
* *

The subject of Art
The materia prima

Bruno de Lanzac: *"The essence in which the spirit we seek dwells has entered and engraved in it, although with imperfect traits and lineaments; the same thing is said by Ripleus Anglois at the beginning of his Twelve Gates; and Ægidius of Vadis, in his Dialogue of Nature, shows clearly and as if in letters of gold there remained, in this world, a portion of this first Chaos, known, but despised by everyone, and which is sold publicly, the same author still says that this subject is found in several places and in each of the three kingdoms; but if we look at the possibilities of nature, it is certain that metallic nature alone must be helped by nature and by nature; It is therefore only in the mineral kingdom, where the metallic seed resides, that we must seek the subject proper to our art."*

...We must take care that the metallic essence is not only in action, but also in potential. It is certain that metallic nature alone must be helped by nature and by nature; It is therefore only in the mineral kingdom, where the metallic seed resides, that we must seek the subject proper to our art."

The revelations of the Adept who veiled his name under the initials of B.D.L. are formal, the subject of art is metallic, it must be aided by another subject of the same nature.

He took many precautions, Bruno de Lanzac, to avoid divulging the name of the secret raw material, so many precautions around an open secret.

The mineral and metallic raw material has been perfectly known for millennia, often used in a metallic state for the purification of gold by cupellation, or in powder form for eye makeup, the Arabs still call it Al

Kohol today, equivalent to the Acohol of the ancients. It is natural, this ore can be easily reduced to metal without grilling.

Before continuing and to avoid the slightest confusion, it is necessary to specify that natural antimony sulfide is often cited as being the possible raw material, the old dragon, but there is another candidate of which Fulcanelli revealed the initial letter, this is the G of Freemasonry, and which is the true Kohl of the Arabs, Galena!.

Philalethes: *"I took part of the igneous dragon and two parts of the magnetic body. I prepared them together by a wheel fire, and by the fifth preparation about eight ounces of true philosophical arsenic were made."*

Philalethes: *"It is a chaos or a spirit, because our igneous dragon, although it overcomes everything, is nevertheless penetrated by the odor of vegetable saturnia. By the union which takes place between his blood and the juice of Saturnia, an admirable body is formed, which is nevertheless not body, because it is entirely volatile, and is not also spirit, because It looks like metal melted in fire. It is therefore effectively a chaos, which is with respect to all metals like their mother, because I know how to extract and draw all things from it and, even, I know how to transmute the sun and the moon through it without the elixir. And whoever has seen it as I have can bear witness to it."*

We call this chaos our arsenic, our air, our moon, our magnet, our steel, however under various considerations, because our matter passes through various states [and undergoes various changes], before the royal diadem is taken from the menstruation of our prostitute. »

Philalethes: *"to resolve the difficulty, read carefully: We must take from our igneous dragon, which hides magic steel in its belly, four parts; of our magnet, nine parts. Mix them together with a burning fire in the form of mineral water, above which will float a foam that you will set aside. Leave the shell and take the core. Purge it and cleanse it three times with fire and salt; and this will be done easily, if Saturn has seen and considered its beauty in the mirror of Mars."*

It is with antimony, therefore stibnite, that Artéphuis begins his presentation: *"Antimony is parts of Saturn, having in all its ways its nature, thus this saturnine antimony is suitable for the sun having silver in itself. -quick in which no metal submerges except gold: that is to say only truly the sun submerges itself in the saturnine antimonial quicksilver, without which quicksilver no metal can be whitened. It therefore whitens the brass, that is to say the gold, and reduces the perfect body to its first nature."*

We must be perceptive, because here the antimony of the wise is perhaps Old Saturn, like all alchemists, Artephius uses an old artifice to veil the vulgar name of the chosen mineral.

Saturn, in the alchemical sense, is the father of metals, it is also the anagram of Natures, Fulcanelli tells us who continues elsewhere, *"As for the gross subject of the work, some call it Magnesia lunarii; others, more sincere, call it Lead of the Sages, Vegetable Saturnia. Philalethes, Basil Valentin, the Cosmopolitan call him the Son or Child of Saturn. In these various names, they sometimes consider its magnetizing and attractive property of sulfure, sometimes its meltable quality, its easy liquefaction. For everyone, it is the Holy Land (Terra sancta); finally, this mineral has as its celestial hieroglyph the astronomical sign of Ram (Ariès)."*

Antimony is the eldest son of Saturn, Flamel tells us, it's the voracious gray wolf of Basil Valentin *"This is why, if you want to operate through our bodies, take the gray and greedy wolf who, by his name, is subject to the warlike Mars, but, by his birth, is a child of old Saturn, throw him the body of the king, so that he may find his substance there, and when he has swallowed up the king, make a big fire and throw the wolf into it, so that he is completely consumed there, so that the king may be deliver*ed again. When this is accomplished three times the lion has overcome the wolf, and he will no longer find anything to eat and our body is then perfect for the beginning of our work.", elsewhere the Adept specifies: "Antimony is the bastard of lead just as bismuth, or marcasite, is the bastard of tin. Antimony is placed between tin and lead, while bismuth and magnesia are placed between tin and iron...

Michaël Maier: « *The artists have their antimony, different from that of the vulgar, although they do not reject the starmade martial regulation that is obtained from the latter, but they use it for different uses."*

"*We could not describe our vitriolated egg more clearly, provided that we know one of the children of Saturn, namely, Triumphant Antimony,*" writes an anonymous author.

Jacques Toll: "*And you finally, whoever you are, and who still doubt what I tell you, just melt some Antimony, and apply yourself to see exactly what is happening; you will see all these things there, you will see the Doves of Philalethes, you will hear the song of the Swans of Basil, and you will see this Sea of the Philosophers, which I explained at greater length in my treatise on fortuitous and unforeseen events.*"

The Cosmopolitan: "*It is a stone and not a stone: it is called stone by its resemblance; firstly, because its mining is truly stone at the beginning that it is taken out of the caverns of the earth, it is a hard and dry material which can be reduced into small parts and which can be crushed like a stone. Secondly because after its destruction which is only a stinking sulfure which must first be removed... And to speak more clearly, it is our magnet, otherwise our steel; and in this sense Hermes wants his father to be the Sun, and his mother the Moon, and that the Wind carried him in his womb. Vulgar air generates or makes appear this magnet, and this magnet generates or feeds the Air or Mercury of the Philosophers, which is the son of the Sun, of the Moon, because it is drawn from the rays of the Sun and of the Moon by the force and attractive virtue of this physical magnet, or this magnetic steel, which is found in every place and at all times; and it is this alkali salt that the Philosophers called armoniac and vegetable salt, hidden in the belly of magnesia. It is called magnesia, because, by a magnetic and occult virtue, it attracts to itself the son of the Sun at the same moment that it takes on its current existence.*"

Flamel: "*You will first consider taking the eldest of the first son, child of Saturn, who is not the vulgal 9 parts, of the chalybe sword of the warrior god 4 parts. Make them redden in a crucible, when they have melted red, throw the 9 of Saturn that I told you, in, then this one will suddenly eat the other: clean beautifully of the fecal garbage coming to the mount of Saturnia with saltpeter and tartar four or five times that will be good when you see a star sign on it regulates it in star mode."*

Le Breton: *"There is a mineral known to true scholars hidden in their writings under various names, which abundantly contains the fixed and the volatile."*

Limojon de Saint Didier: *"Our stone is born from the destruction of two bodies, of these two bodies, one is mineral, the other metallic and both grow in the same earth."*

According to this extract from the works of Mr. Grimaldy, Head of the University of Medicine of Chambery, it is easy to understand the importance of the alchemical mineral, the sole dispenser of Mercury. *"Let us move on to the choice of Antimony, & the various names given to it by those who wanted to hide its preparation & mysteries, so that it would serve to understand their enigmas, & for the explanation of their hieroglyphs...."*

"The Philosopher Chemists depict this mineral to us with a character which represents the world with the cross above, to point out to us that as the mystery of the cross purifies & saves the soul from all its spiritual stains, Antimony & its remedies well & gently prepared, purify & deliver the body from all the impurities that cause & maintain the illnesses that afflict it. They call it by several enigmatic names like the Wolf, because it consumes and devours all metals, with the exception of gold. Others have called him Protheus, because he takes on all kinds of forms, and because he is clothed in all colors by means of fire. Others call it the root of metals, both because their roots are found close to it, and because there are some who believe that it is the root and principle of metals. Or it is also called Sacred Lead, Lead of the Philosophers, and Lead of the Sages, because it has some connection to the nature of Saturn which devoured its children as it devours metals, & because there are some who take it for the subject of the Great

Work of the Philosophers, & their quintessence. Glauber describes it to us as the first being of gold."

But the great Adept Fulcanelli does not seem to share this opinion:

"The most educated of our people in the traditional cabal were undoubtedly struck by the relationship existing between the path, the path traced by the hieroglyph which takes the form of the number 4, and the mineral antimony, or stibium, clearly indicated under this topographic term. Indeed, natural antimony oxysulfide was called, among the Greeks, Stimmi or Stidi, or Stidia is the path, the path, the path that the investigator (Stideyj) or pilgrim travels on his journey; it is she whom he tramples underfoot (Steidv). These considerations, based on an exact correspondence of words, did not escape the old masters nor the modern philosophers, who, by supporting them with their authority, contributed to spreading this error that common antimony was the mysterious subject of the art... We know that the alchemists of the 14th century called Kohl or Kohol their Universal Medicine, Arabic words al cohol, which mean subtle powder, a term which later took on in our language the meaning of brandy (alcohol). In Arabic, Kohl is said to be pulverized antimony oxysulfide, which Arab women used to dye their eyebrows black... We would be of the same opinion, if we did not know that not the slightest molecule of stibnite in the platyophthalmon of the Greeks (sublimated mercury sulphide), the Kohl of the Arabs and the Cohol or Cohel of the Turks. The last two, in fact, were obtained by calcination of a mixture of peened tin and gall nuts, such is the chemical composition of the Kohl of oriental women, which ancient alchemists used as a term of comparison to employ the secret preparation of their antimony."

Indeed, antimony, when it comes to the mineral known as stibnite, is not the subject of the Great Work. But Fulcanelli intentionally or not makes an error when he indicates the composition of Kohl!

Kohl, in Arabic, actually designates galena, a natural lead sulphide, finely powdered, used for millennia by Sumerian women, then by the Hebrews, then by the Egyptians to make up their eyelids and eyebrows. It is not a mixture of peened pewter and gall nuts, as Fulcanelli would have us believe. Fulcanelli must necessarily have known the meaning of the Arabic word "*Kohl*" and the exact nature of the makeup that oriental women use to make up their cosmetics. ! The word Kohl comes from a root of three consonants which means black in Arabic, and Al kohl, a masculine word, means black. Kohl, a black ore reduced to an impalpable powder used by Arab women among others, clearly describes the true subject of the wise men and the way of reducing it to an ultra-fine powder.

Many times I have had the kohl in my hands from the time when, as a teenager in this small village of Aurès to the south-east of Algiers, I had fun drawing with the little stylus on the pages of my school notebooks not without having forgotten to moisten the fine black powder to make it suitable for impregnating the frame of the paper;

"*I am beautiful and I am black*," King Solomon says about art in the Song of Songs.

It is the word Al kohl, taken from Arabic, which gave rise in Europe to the word alcohol and then alcohol, the initial meaning of which has always been associated with the mysterious material necessary for making stone. He would later designate, but wrongly, alcohol, a product obtained by the distillation of wine.

Kohl, galena in a powdered state, as I was able to observe during recent travels, is still used today by women in North Africa and the Middle East. In Morocco, in the souks of Casablanca or Marakech it is possible to obtain an elegant wooden tube of kohl with its stylus which acts as a lid, for a handful of dirham.

Fulcanelli, an expert at blurring the lines, suggested that the letter G was the initial of the vulgar name of the subject of the wise men. G is also the initial of Gea or Gaia, the Earth, our alchemical Earth. In bird language anything goes, including passing bladders off as lanterns. Fulcanelli here makes a nod to his Free Mason brothers, for whom the letter G has an initiatory importance.

Galena versus Stibine?
Or Galena or Stibine?
I vote for Stibine!

The Athanor

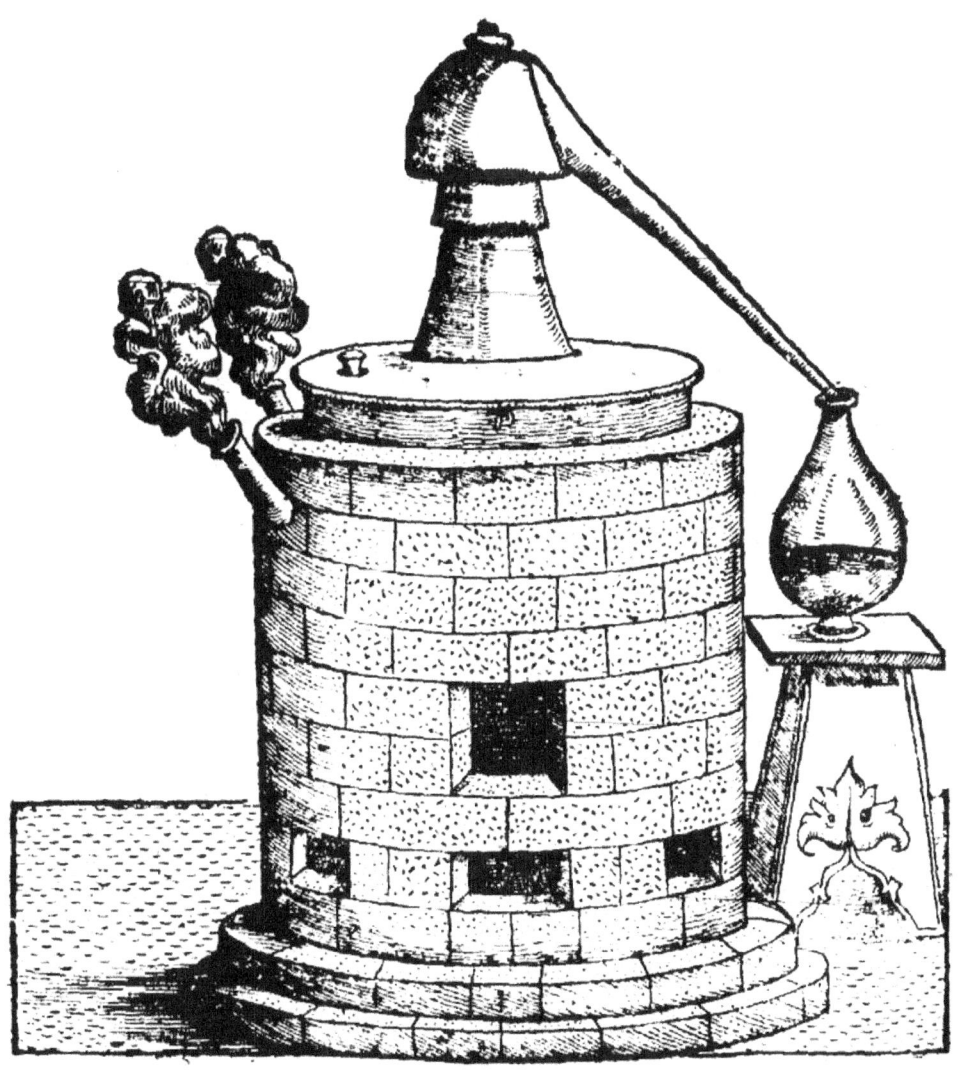

The Alchemist oven and its Aalambic

Mars, the Vaillant Knight, the giver of Sulfure

The raw material being known, there is only one step to take to discover its protagonist.

Philalethes: *"As steel pulls the magnet towards itself, so the magnet turns towards the steel. This is what the Sages' magnet does to their steel. This is why having already said that our steel is the mine of gold, it must also be noted that our magnet is the true mine of the steel of the Sages."*

Cyliani: *"Don't forget that the mysterious solution of matter, or the magical marriage of Venus with Mars, took place in the temple of which I previously spoke to you, on a beautiful night, the sky calm and without clouds, and the sun being in the sign of Gemini, the moon being in its first quarter at its full, with the help of the magnet which attracts the astral spirit of the sky, which is seven times rectified until it can calcine the gold."*

Basil Valentin: *"By the addition of tartar and salt, we make, with antimony, a repel, which, being melted, if we add steel by a secret preparation, it becomes starry, which was before me called the star of the wise. If sometimes it is melted with saltpetre, it becomes yellow with an igneous property."*

Fulcanelli did not hesitate to designate iron as the true dispenser of sulfure which should not be confused with the chemical element S, alchemical sulfure is the second principle. For Fulcanelli, the male or fixed protagonist appears disguised under the mask of Mars, the Knight or the Saint slaying the Dragon, (Saint-Marcel, Saint-Michel, Saint-George, Sainte-Marthe). He is the King or the Lion of Basile Valentin, in the fight symbolizing the preparation of the first material.

Iron is the Knight Ares, Fulcanelli tells us again. Etymologically, Ares is what gives form to individuals. It is our Magnet which attracts our Magnesia whose astrological symbol is the ram – the stibnite or Aries – which is why he recommends differentiating Arles, Arès or Aries. The magnet is iron, taken as a lover, it is he who has this affinity for the sulfure (S) contained in the stibnite, the magnesia that it will attract as the magnet attracts iron, simple simple chemical reduction reaction.

Iron is also the mining of gold by the learned Lintaut and the Presiden d'Espagnet, whose expression Fulcanelli uses in his two books, where he devotes numerous chapters to the base metal and reviews the different properties that adepts and spagyrists attribute to it. He also mentions iron which emerges from the motto of the great financier Jacques Cœur, – A vaillant cuers riens impossible –, iron, which is the common name for the basic material worked, writes the Master.

Elsewhere, analyzing the hieroglyph of the Griffin, in the fight of the "*knight armed with lance and shield, and the old arsenical dragon*", he describes the process by analogy with the passion of Christ; "crucify with three iron points, so that the body dies and can then be resurrected." Further on, commenting on the book of Abraham the Jew, the Master once again underlines the operational detail: "it is," he says speaking of the old *grimoire, written with an iron point whose secret property changes the intimate nature of our Magnesia.*"

All these comments on iron are very relevant. The metallurgist is well aware of the properties of iron and its usefulness in the reduction of antimony ore until the star (compos stellae) is obtained. The choice of iron as antagonist also confirms the choice of stibnite, the primitive mercury which, in close association with iron, the only provider of sulfure, will give rise to Rebis.

We will once again notice the confusion maintained about stibnite, the only mineral to be cited in the grimoires, and about antimony, the regulin part obtained after the attack on the ore with iron.

The Mediator, the Third Agent, the Salt, the Secret Fire, the Water of the Wises

We have the old arsenical dragon (the stibnite or galena), the knight armed with the lance and the shield (iron), all we have to do is find the Mediator.

Salt, double according to Eugène Canseliet, is one of the essential elements of the Great Work. However, we must be careful not to take Canseliet's assertions at face value.

Salt is represented by a priest celebrating the union of the King and Queen, on the sixth key of Basil Valentin. Canseliet revealed that another symbolic representation is often found in the appearance of little angels (parvuli). Fulcanelli considers salt when he evokes the mythical Eros-Cupid.

For Fulcanelli: *"Of these principles, two are reputed to be simple, sulfure and mercury, because they are found naturally combined in the body of metals; only one, salt, appears to consist partly of fixed substance, partly of volatile matter. We know, in chemistry, that salts, formed of an acid and a base, raise, through their decomposition, the volatility of one, as well as the fixity of the other. As salt participates both in the mercurial principle through its cold and volatile humidity (air), and in the sulphurous principle through its igneous and fixed dryness (fire), it therefore serves as a mediator between the sulfure and mercury components of our embryo."*

De Pontanus: *"All our application and care, therefore, must be only to the knowledge of true practice, in the first, the second, and the third work. It is not the fire of bath, of dung, nor of ashes, nor any of all the other fires that the Philosophers sing to us, and describe to us in their books.*

What then is this fire which perfects and completes the whole work, from the beginning to the end? Certainly all Philosophers have hidden it; but, for my part, not being bound by any oath, I want to declare it with the entire accomplishment of all the work.

The Philosophers' Stone is one, and one, but hidden and enveloped under the multiplicity of different names, and before you can know it you will take great pains; you will hardly find it of your own genius. It is aqueous, aerial, igneous, terrestrial, phlegmatic, angry, sanguine and melancholic. She is a sulfur and likewise Quicksilver.

It has several superfluities, which, I assure you by the living God, are converted into true and unique Essence, through our fire. And he who separates something from the subject – believing it necessary – certainly knows nothing about Philosophy. Because the superfluous, the dirty, the filthy, the ugly, the muddy, and generally all the substance of the subject, is perfected into a fixed spiritual body, by means of our fire. What the Sages have never revealed, and makes few people achieve this Art; imagining that something dirty and naughty must be separated.

Now we must reveal and draw out the properties of our fire; if it suits our matter according to the manner in which I spoke, that is to say if it is transmuted with matter. This fire does not burn matter, it does not separate anything from matter, does not divide or separate the pure parts from the impure, as all Philosophers say, but converts the whole subject into purity. He does not sublimate like Géber does sublimations, and Arnaud similarly, and all the others who have spoken of sublimations and distillations. It is made and perfect in a short time.

This fire is mineral, equal and continuous, it does not evaporate, unless it is too excited; it participates in sulfur, it is taken and comes from somewhere other than matter. He breaks, dissolves, and congeals all things, and likewise congeals and calcines; it is difficult to find by industry and by Art. This fire is the summary and shortcut of the whole work, without taking anything else, at least a little, and this same fire is introduced and is of poor ignition; because with this little fire the whole work is perfect, and all the required and due sublimations are made together.

Those who read Géber and all the other Philosophers, when they live a hundred million years, will not be able to understand it; because this fire can only be discovered through profound meditation of thought, then it will be understood in books, and not otherwise. The error in this Art consists only in the acquisition of this fire, which converts matter into the Philosophers' Stone... So apply yourself to this fire, because if I myself had first found it, I would not have wandered two hundred times on the actual matter. Because of which I am no longer surprised if so many people cannot achieve the accomplishment of the work."

Canseliet quotes a passage from the Rosary of the Philosophers: *"The salts and alums are not the stone, but the helpers of the stone. He who has not tasted the flavor of salts will never come to the desired ferment of ferment; because it ferments the defined by excellence.*
The superior is like the inferior.
Burns in water, washes in fire.
Cook and recook, and cook again.
Very often dissolve and constantly coagulate.
Kill the quick and resurrect the dead.
And that seven times. And you will really have what you are looking for,
If you know the regime of fire. Mercury and fire are enough for you."

Nicolas Flamel: *"We will first consider taking the eldest of the first son, child of Saturn, who is not the vulgal 9 parts, of the chalybé saber of the warrior god 4 parts. Make them redden in a crucible, when they have melted red, throw the 9 of Saturn that I told you into, then this one will suddenly eat the other: clean beautifully from the fecal garbage coming to the mount of Saturnia with saltpeter and tartar four or five times that will be good when you see a star sign on it regulates it in star mode. Therefore from gold is made the key and cutlass which opens and incises all metal even especially Gold, Silver and Mercury all which eats and devours and keeps in its belly, and through it a right machine, path of truth appears, if as worked as is profession, because this saturnal machine is the triumphant regal herb for what it is*

Money and little imperfect king that we promote to the degree of much glory and honor and is even the queen, that is to say the moon and wife of the sun."

Canseliet: *"The saltpetre of the philosophers, according to the undeniable etymology of its name, Latin and consonant, designates the salt which is for the stone or even which belongs to the stone: Sal petræ,"*

Then, commenting on an extract from *"Three Books of the Potier"*, by Cyprian Piccolpassy cites this passage: *"It is to be known that with us the lees of the wine are collected more in the months of November and December, than not in other times, seeing that it can be collected tartar, since the tons are well dried, means those where the wines have remained for a long time. These are shaved inside with an iron, a bad crust will come off with one or two fingers, that's the tartar…"*

"We will add the certainly not insignificant advice, not to use any artifice in order to make the solution of tartar easier. The dual factor of time and patience, in this case, plays a big role,…"

Huginus a Barmâ: *"Where is the water of the Sages found? Hermès writes: We draw our water from a sordid and stinking menstruation and Dantin: Our water is found in old stables, latrines, cesspools; know that fools do not hear these words, they believe that this is mercury; note, however, that it is not mercury that the philosophers are talking about but a dry water which brings together all the mineral spirits, the soul and the body, by making them penetrating, and which after having brought them together abandons them, separates them from them and leaves them in the state of fixity."*

Bernard Husson: *"The caregivers, through their close union with the alchemical couple, transform the appearance and functions of its protagonists to such an extent that the best authors use the terms of reversal to describe the operation they provoke. or repayment. Nothing better accounts for these sudden phenomena than the sudden irruption on the stage of the chemical theater of unforeseen actors, under whose new costume and mask the neophyte will have to recognize one of the old characters (or several melted into one). . This is why, if four chemically defined bodies, in the current sense of the term, each playing the role of one of the traditional cosmic elements, are actually required for the start of the operation mentioned in this first key, we should not believe that each one corresponds unequivocally to one of the four figures on the board."*

Limojon de Saint-Didier: *"All you can reasonably expect from me is to tell you that the natural fire, of which this Philosopher speaks, is a potential fire, which does not burn the hands; but which makes its effectiveness appear as long as it is excited by the external fire. It is therefore a truly secret fire that this Author calls Lunatic Vulcan in the title of his story. Artephius gave a more extensive description of it than any other philosopher. Pontanus copied it and showed that he had erred two hundred times because he did not know this fire, before he had read and understood Artephius: this mysterious fire is natural, because it is of the same nature as philosophical matter; the artist nevertheless prepares both... Consider only with application that this natural fire is nevertheless an artful invention of the artist, that it is capable of calcining, dissolving and sublimating the stone of the Philosophers and that there is only this one kind of fire in the world capable of producing such an effect. Consider that this fire is of the nature of lime and that it is in no way foreign to the subject of Philosophy. Finally consider by what means Geber teaches to make the sublimations required for this art: for me I can do no more than to make for us the same wish that another Philosopher made: Sydera Veneris et Corniculatae Dianae tibi propitia funto."*

The Cosmopolitan: *"Take what is, but which cannot be seen, until it pleases the artist; it is the water of our dew, from which the saltpeter of the philosophers is drawn, by which all things grow and are nourished."*

From Grimaldi: *"Agreeing that everything sublime that the Philosophers say about Nitre is true, we must at the same time agree that they hear of an aerial Nitre, which is drawn into salt whiter than snow, by the force of the rays of the Sun and the Moon, by a magnet which attracts the invisible spirit; this is the magnesia of the Philosophers."*

Fulcanelli: *"So if you wish to possess the griffin – which is our astral stone – take two parts of virgin earth, our scaly dragon, and one of igneous silver, which is this vaillant knight armed with the lance and the shield. Ares more vigorous than Aries must be in less quantity. Pulverize and add the fifteenth part of this pure, white, admirable salt, washed and crystallized several times, which you must necessarily know. Mix thoroughly; then taking the example of the passion of Our Lord, crucify with three iron points, so that the body dies and can then be resurrected. Once this is done, remove the coarsest sediment from the corpse, crush and crush the bones; mix everything over a low heat with a steel rod. then throw into this mixture half of the second salt taken from the dew which, in the month of May, fertilizes the earth, and you will obtain a lighter body than the previous one. Repeat the same technique three times, you will reach the mining of our mercury, and you will have climbed the first step of the staircase of the Sages."*

Basil Valentin: *"Everything reduced to ashes shows and manifests its salt. If, in dissolution, you know how to keep separately its sulfur and its mercury, and of these two to render with industry what is necessary for the salt, it will be possible to make the same body as before its dissolution... On the day of the Last Judgment , the world will be judged by fire, and that which was made from nothing by the Master, will be reduced by fire again to ashes, from which the Phoenix will produce its young, because in these ashes the true tartar is hidden, and after its dissolution, the strong lock of the Royal palace can be opened.*

The Master who has no ashes cannot prepare salt for our art either, because without ashes our work cannot be corporified, because salt is the only one that hardens all thing.

As salt preserves all things, and preserves them from decay, so the salt of our masters defends and preserves metals and prevents them from being entirely destroyed, by preserving their balm and their spirit, for otherwise there would remain a dead body, from which something fruitful could no longer be produced, because the metallic spirits would have left it, and their spiritual balm and salt incorporated in nature, perishing, the body will be dead."

Monte-Snyder: *"Tartar improves metals, makes them malleable, and this is why it agrees with metallic salt, which, through tartar, is multiplied... Just as the earth is in the presence of air, so Salt is also found opposite the spirit; and, again, just as air is a mediator between fire and earth, not differently the Ruah of the sages is the mediator between the body and the soul, the true knot of what unites the soul with the body; and this intermediary of which I speak, is to be compared to some double igneous man: in the great philosophical work, he is double, and, moreover, by obtaining his igneous vivacity, hermaphrodite, since life consists in fire , who, of no one, if not only of himself, lives and moves, and so on.*

But in order to access the subject and my goal, know then that the mineral and metallic fire, in itself and for itself, is raw material, which is found in the mineral of Saturn, or in its receptacle, or Universal home: from this universal home, which is his, he must, from time to time, retreat, because of the affliction of the igneous flying dragon.."

Canseliet: *"Under the effect of heat applied wisely, thanks to its subtle nitre, dew elevates and ennobles any salt whatsoever and, preferably, those that nature has reserved for the Great Work. In the company of this pair of salt melts, the nocturnal condensation undergoes the action of fire easily and without damage.*

"What is this white salt that we must use, preferably, crystallized into snow, and which is easily mixed with our mineral and our metal, themselves divided, one into powder, the other into filings?

Our salt or, if you prefer, our fondant, is double because it is physically composed of the ana addition of two different salts. "

The reader understands that two salts must be used, the term ana, used by Canseliet, meaning in equal proportions. But alas, gross errors are hidden beneath these apparently innocuous explanations; I will come back to this later.

The first salt, nitre salt, or saltpeter, with the chemical formula NO3K, in other words potassium nitrate, was obtained by brushing old walls and plasterwork, it was then given the name of covering saltpeter. Niter should not be confused with the Egyptian natron. Potatasium nitrate had been known to the Chinese for thousands of years, it was used in the making of rockets and later explosives.

The second salt, tartar, actually refers to potassium acid tartrate or bitartrate, a natural product collected during wine fermentation. The term scale today more generally refers to certain saline deposits, such as calcium carbonate in water pipes for example.

Tartar plays a secondary role at the very beginning of the work and it may be possible to exclude it from the process by dry method.

The Dew

We are approaching, here, the mysterious and magical domain of alchemy, and it is undoubtedly the use of this agent which has given it its astrological and mystical connotation, taking into account the conditions of its capture, at night, at night. equinoxes, the moon in the ascending phase.

Unlike the other protagonists of the Great Work whose trace can be traced over a long period, dating back to Mary the Jew, the rare allusions to dew only appear towards the middle of the Middle Ages. The story of the dew perhaps begins with the Romance of the Rose, by Jehan de Meung. No one can affirm that followers like Artéphuis, Flamel, le Trévisan, Arnaud de Villeneuve, Basile Valentin or Philalèthe, actually implemented the dew.

On the other hand, it is indeed the dew that Jacob Sulat, alias Altus, uses in his Mutus Liber, on the fourth plate, reproduced below, as does Cyliani, in his Hermès Dévoilé, as does the anonymous author Scholies and Hermetic Recreations from which Cyliani was most certainly inspired.

The plates of Mutus Liber are very telling, despite the silent qualifier used by the author. We will notice on the fourth plate, the presence of the ram and the bull symbolizing the two spring months, the cosmic radiation, in the form of a fan, giving its mystical character to the operation, and the triangular arrangement of the sheets stretched on stakes, discreetly highlighting a certain analogy with the secret fire.

The Cosmopolitan specifies: *"Take what is, but which cannot be seen, until it pleases the artist; it is the water of our dew, from which the saltpeter of the philosophers is drawn, by which all things grow and are nourished."*

It is also the dew that Limojon de Saint-Didier uses in his commentary on the "*Ancient War of Knights*", and that the nymph of Cyliani declares to be the essential vehicle.

Limojon de Saint-Didier: *"As the wise man undertakes to do something through our art, which is above the ordinary forces of nature, like softening a stone, and making a metallic germ vegetate, he finds himself inevitably obliged to to enter through deep meditation into the most secret interior of nature, and to make prevail the simple but effective means that it provides; now you must not ignore that nature from the beginning of Spring, to renew itself, and put all the seeds, which are within the earth, in the movement which is proper to vegetation, permeates all the air which surrounds the earth, of a mobile & fermentative spirit, which takes its origin from the father of nature; it is properly a subtle nitre, which creates the fertility of the earth of which it is the soul, and which the Cosmopolitan calls the salt-peter of philosophers."*

Further, perhaps assimilating dew to secret fire: *"All you can reasonably expect from me is to tell you that the natural fire, of which this Philosopher speaks, is a potential fire, which does not burn hands*

; but which makes its effectiveness appear as long as it is excited by the external fire. It is therefore a truly secret fire that this Author calls Lunatic Vulcan in the title of his story."

We will notice the Lunatic character attributed to Vulcan.

The author of Hermetic Recreations, a manuscript discovered by the hermetic scholar Bernard Husson, is the first to highlight the polarizing power of the moon in scientific terms: *"Everyone knows today that the light that the moon sends us n It is only a borrowing from that of the sun, with which the light of the other stars is mixed. The moon is therefore the common receptacle or focus of which all philosophers have heard; she is the source of their living water. So if you want to reduce the sun's rays to water, choose the moment when the moon transmits them to us abundantly, that is to say when it is approaching its fullness; By this means you will have the igneous water of the rays of the sun and the moon in its greatest force."*

We have just reviewed the most critical points of this Art, affecting the raw materials as well as their properties.

Thanks to our pioneering scholars, Canseliet, Husson or Atorène, it was not very difficult to reveal these secrets so coveted over the past two millennia.

We therefore have not only a precise and concrete idea of the secret materials used by the adepts, but also an overview of the techniques implemented.

*
* *

The Philosopher's Stone in 5 points

The process consists of manufacturing a gem with extraordinary physico-chemical properties, unknown to modern physics, the Philosopher's Stone.

This gem, which appears as crushed glass with a color tending to dark garnet red, has an inexplicably high density, perhaps twenty to thirty times that of water. Fuse at low temperature, it is very soluble in water or alcohol, when it is not oriented towards transmutation, and in this state becomes the famous elixir of youth (to be consumed with great moderation).

In its undifferentiated state, it can be both multiplied in quality and quantity to infinity. To do this, it is enough to resume the process from the regime of the Eagles or the sublimations, with a new portion of Philosophical Mercury, and to repeat the great concoction. Each phase of increase and multiplication will only require a fraction of the time of the previous one, approximately an eighth, according to Fulcanelli, half according to other authors. The Stone will also see its power increase tenfold with each multiplication and its fusibility increase significantly to the point of becoming liquid then gaseous, if the limit number of reiterations is exceeded. In this state, it radiates a strong light, and becomes the very mysterious perpetual lamp.

To direct the Stone towards metallic transmutation, it will be enough to ferment the Stone with four to ten times its weight of gold in a crucible over open fire for a few hours. In this state it will have no effect on the other two kingdoms.

Depending on the degree of multiplication, its tinctorial power may vary from a few hundred to several thousand times the metallic masses involved. The tinctorial power of the Stone depends on the nature of the metal used during the projection. Transmutational potencies of 1:20,000 have been reported.

Finally, during most of the transmutations recounted and commented on by Bernard Husson in Alchemical Transmutations, unexplained violations of the principle of conservation of mass were observed, endowing the Philosopher's Stone with surprising properties such as the increase in the mass of metal, as well as the ability, for synthesized gold, to further transmute the initial base metal by inquartation and fusion.

In short, the dry process, detailed below, can be summarized as follows:

At the very beginning, the alchemist must select the canonical materials, revealed above, and subject them to the very first preparations which are the responsibility of physical chemistry or metallurgy.

From the above, we will conclude that, in all, four bodies are required, i.e. two salts, a mineral and a metal.

An axiom explains that the Work neither admits nor receives what comes from the outside. This means that nothing foreign should be added to the Stone, once the process has begun.

The preliminaries carried out:

1 - The first operation consists of obtaining the Astral Stone or the Griffon. This operation is also called Separation, because it is necessary to strike a sharp blow on the cooled crucible after the fire test, in order to detach the regulin part (refined metal) to separate it from the slag. It is also the pilgrimage to Compostela that the Artist metaphorically accomplishes in order to obtain the starry compost, according to Flamel and Fulcanelli.

2 - The second operation, which alchemists often call the first, because the most important and the most secret, leads to the development of the mercury of the wise, at the end of a long series of manipulations called philosophical sublimations and which It

should not be confused with the modern chemical operation. Starting from the materials obtained in the first work, the alchemist carries out a liquid-solid extraction to obtain a substrate, symbolically called the Remora, the King, the bean, the Rebis, etc., which he recovers by skimming the surface of the compost supercooled. This operation is also called Eagles, by analogy with the raptor which removes its prey, that is to say the volatile which removes the fixed. The motif of our cover well illustrates this operation, considered the most important of the magisterium: the caduceus of Hermes – a winged helmet surmounting a scepter around which two serpents coil – is linked to a marine anchor. The composition is bordered by two cornucopias, attesting to the importance of the final result. The two spears, crossed to emphasize the need to work in the crucible of the dry method, are decorated with a scallop shell, the usual representation of the philosopher's mercury. an important axiom relates to this operation: "If you know how to dissolve the fixed and fix the volatile, you have reason to console yourself."

Dissolving the fixed and coagulating the volatile, that's our entire philosophy, Fulcanelli and the great Masters repeat endlessly.

A clarification, the volatile which is mentioned in the dry method is not a material which can be distilled, it is a solid body at room temperature and liquefiable in the crucible.

We must at all costs avoid sophist paths, such as those described for example on the internet by false alchemists who feel elevated to the level of the greatest masters without even having passed the stage of learning or a healthy initiation.

3 - The third operation, also called Coction, consists of subjecting matter to the ultimate igneous transformation. Fulcanelli, usually more verbose, was extremely reserved about this phase and the scattered teaching he gives through his trilogy is difficult to follow. This operation makes it possible to obtain sulfur, the last stage of the transformation, which will only require gradation to increase its power. Coction requires more skill than all the other phases.

You must take the Rémora from the second work, without removing it from the environment where it is located, that is to say immersed in its salt, then make it undergo a final cooking in a crucible over an open fire for 4 to 6 days. When the time has expired, the crucible splits and reveals the crimson carbuncle in a red, blistered and opaque case, similar in shape to a chestnut or a hedgehog.

4 - The fourth alchemical operation is called multiplication. It consists of increasing the weight and tinctorial power of the Stone. According to ancient adepts, the properties of the Stone improve gradually, with each multiplication by a factor of ten, while the time required for each new imbibition decreases by a proportion of one eighth of the previous time. Finally, at a certain stage of multiplication, the stone can no longer solidify, it becomes liquid and very fluorescent, operations must then be suspended otherwise it will be lost during the transition to the gaseous state.

5 - The last operation is called fermentation and consists of directing the Stone towards transmutation. Once the multiplications in weight and power are completed, the Stone is simply melted with three or four times its weight in gold. Oriented towards projection, the Stone has no effect on the other kingdoms. Unfermented, it remains susceptible to being increased in weight and can be used as medicine in certain conditions. Alchemists have the habit of reserving part of the Stone before fermentation in order to use it for medicinal purposes.

Before approaching, in detail and in the right order, the alchemical techniques, which to my knowledge has never been done to date, I would simply like to give my feelings on a fundamental point in the process of the Great Work: dew.

According to Eugène Canseliet, thanks to its subtle nitre, dew ennobles all salt and particularly those reserved for the Great Work. By dissolving the binary salt in the dew and then distilling the mixture thus obtained, we would harvest a substrate enriched with "exalted" celestial nitre and endowed with new properties, essential for the elaboration of the Stone.

I do not share this opinion, dew seems absent from the processes implemented by many followers, although Canseliet, as well as other modern authors, maintains that without dew it is impossible to collect the green-colored salt during the first calcinations. Well no, it is entirely possible to obtain the green color under certain conditions from the first work, without requiring the use of salt treated with dew.

It therefore remains to determine the necessity of these long series of painful operations which involve the dew and which seriously complicate the alchemical process.

To cut this Gordian knot, we could, for example, carry out two series of manipulations at once, one with and the other without the use of salt treated with dew, all other conditions being strictly identical, it will be relatively simple to rule on this point of doctrine.

Arcade of Nicolas Flamel at the portal of Saint-Jacques-la-Boucherie-destroyed in 1790.

Plate 4 of the Mutus Liber - Harvest of dew.

Dry Way Process

The initiated alchemist has the necessary materials and knows their properties:
— the black virgin, the mythical dragon, our stibnite (galena will not be considered here), in its raw state, the sole dispenser of mercury,
— Mars, the vaillant knight, the iron filings, our sulfur,
— Potassium bitartrate, our tartar,
— Saltpeter,

He now needs:
— Equip the laboratory,
— Make a good supply in dew,
— Establish a protocol as complete as the procedures to be implemented, operating conditions, quantities and / or proportions of reagents, duration of operations,
— Define the intermediate steps, the characteristics of the by-products,
— Evaluate potential dangers at each stage (explosion, poisoning).

Before starting the cooking process, you must have carried out the preliminaries that are the work outside the work. These preparations are never mentioned by the authors, they all assume that the apprentice alchemist has the required elements.

Preparation of the ore -Liquefaction
In his summary, Flamel is the first to clearly indicate this process of enrichment outside the work, namely liquation, without omitting to symbolically designate the subject of art:

"From Mercury they believ do
Philosophers & perfect;
But you can never achieve it,
So abuse they lie,
Who is the first material
Stone, & real mining:
But they will never succeed,
Nor will any good be found there,
If they don't go up the mountain
Of the seven, where there is no plain,
And above will look
The fixes from afar he will see;
And above the highest
Montaigne, will know without fail
Royal triumphant herb
Which they named minerale,
called is saturnial,
But leave the mark it is suitable,
And hang the juice that comes from it
Pure and clear: de cecy advises you,
To hear this better;
Because you can do well with her
Most of your business.
This is the real kind Mercury
Very subtle Philosophers."

The great Adept reveals the name of the chosen matter, which is saturnial. This name of triumphant and royal herb is taken from a manuscript attributed to Mary the Prophetess, the term saturniale being borrowed from Artéphuis. This is our stibnite. The Adept recommends taking the pure and clean juice which comes from the mineral Saturnia and leaving the marc (the gangue), in doing so, one will possess the true kind and subtle Mercury.

Stibnite - chemical formula Sb_2S_3 - in its pure state, melts at 550°C and its density is 4.65 kg/dm3. In general, commercial ores have contents between 40 and 70% and, according to Atorène, the acceptable minimum corresponds to around 30%, or a density of 3 kg/dm3 for the poorest ore.

To remove the ore from its gangue, it is melted with an enameler's lamp, in a crucible whose bottom is pierced with a hole of 10 mm in diameter. The stibnite flows at the bottom while the gangue floats the molten sulphide. To do this, Atorène uses a common terracotta flower pot.

Grinding

In order to obtain maximum reactivity from the ore, it must be reduced to an impalpable powder. This operation can be done in a cast iron mortar. To avoid scattering dust, you can proceed in a humid environment by adding water in small quantities.

The mixture obtained will be allowed to settle for two or three hours and the water will be recycled. We will complete this operation before drying the fine powder in the oven, at moderate temperature.

Assation / Cooking

This operation, practiced by Fulcanelli and E. Canseliet, just before drying of the ore subtly crushed in the wet phase, consists in submitting the porridge thus obtained, even diluted with sand, with the low heat, modulating the heating as a function of the 'Lunar activity. The sand will later be eliminated by sieving.

Assation is a spagyric technique that is applied to plants or organic products, extending it to minerals seems very empirical, we can do without this operation.

Preparation of tartar

Potassium acid tartrate or potassium bitartrate is a natural product that is collected in oak barrels or in cement vats that have contained wine.

With formula $C_4H_5KO_6$, it contains a lot of organic impurities which must be eliminated. We proceed by hot dissolution in distilled water or filtered rainwater and by successive crystallizations.

Pure tartar is soluble in 110 g/liter of water at 100°C, and it is poorly soluble when cold (around 10 g/liter at 20°C).

By taking each crystallized fraction and subjecting it to a new solution in clear water, collecting the precipitate in the cooled solution and thus repeating the operation several times, we finally obtain a salt sufficiently pure for our operations. The yield being deplorable, it is necessary to provide at least three times more material than the quantity required in the final.

Some authors recommend purification by calcination, over open fire in a crucible. Although this operation effectively eliminates the lees, there is a risk of decomposition, because from 300°C we finally obtain potassium carbonate with the formula CO_3K, or even, under certain conditions, potassium cyanide.

A question immediately comes to mind.
Can bitartrate or potassium carbonate be used indifferently for the operations of the Great Work?
Not sure. Just know that carbon and hydrogen, both combined in bitartrate, improve the elimination of combustible parts contained in the ore. The molecular masses of the two salts being different, it would seem logical that the proportions used would also be different.

But does this really matter? I think so.

On the other hand, trying to make potassium carbonate from bitartrate is a complete waste.
Indeed, we can more easily extract carbonate from the ashes of most plant species, and this is what Basile Valentin shows in these *"Twelve Keys."*

Collection of dew

Few authors have addressed this subject, apart from Armand Barbault who placed himself on a strictly spagyric level.

Anonymous author, extract from Hermetic Recreations: *"Be assured therefore that without igneous water composed of the pure light of the Sun and the moon, it will be impossible for you to overcome the numerous obstacles which will still multiply before your eyes, when you attempt the passage of this famous Strait which leads to the sea of the wise, this water which some rightly call universal spirit and which the Englishman Dickinson has sufficiently made known, is of such great virtue and penetration that all bodies who are affected by it, easily return to their first being.*

I have already made it known that it is not rain water or dew water which is suitable for this operation, I will add here that it is not water from a species of mushroom commonly called Flos either. Coeli or Flower of the Sky and which we very incorrectly take for the Nostoch of the ancients, but an admirable water drawn by artifice from the rays of the sun and the moon. I will also say that the salts and other magnets that we use to draw moisture from the air are good for nothing in this circumstance and that there is only the only fire of Nature that we can rely on here. serve usefully. This fire contained in the center of all bodies needs a certain movement to acquire this attractive and universal property which is so necessary to you, and there is only one body in the world where it is found with this condition. , but it is so common that it is found wherever man can go; that is why I think it will not be difficult for you to meet him."

The anonymous author has just written that dew or rainwater should not be used, but then states that astral spirits should be harvested under the full moon. This is a typical exposition device put there to embarrass the reader. As if the author, regretting a shameful lie, then wants to correct his presentation, implying that the astral spirit is no longer dew.

"I said that light was the common source, not only of the Elements, but also of everything that exists, and that it is to it, as to its principle, that everything must relate. The Sun and the fixed Stars which send it to us in such profusion are like its generators; but the Moon placed intermediately, soaking it with its humidity, communicates to it a generative virtue by means of which everything is regenerated here below.

Everyone knows today that the light that the moon sends us is only borrowed from that of the Sun, with which the light of the other stars is mixed. The Moon is therefore the receptacle or common focus of which all philosophers have heard: it is the source of their living water. If therefore you want to reduce the rays of the Sun to water, choose the moment when the moon transmits them to us with abundance, that is to say when it is full, or when it approaches its fullness: you will have by this means the igneous water of the rays of the Sun and the Moon in its greatest force.

But there are still certain essential provisions to be fulfilled, without which you would only have clear and useless water.

There is only one time suitable for this harvest of astral spirits. It is the one where Nature regenerates; because at this time the atmosphere is completely filled with the universal spirit. The trees and plants that turn green again, and the animals that indulge the pressing need of generation, make us particularly aware of its benign influence. Spring and autumn are therefore the seasons you should choose for this work; but spring especially is preferable. Summer, because of the excessive heat which expands and chases away this spirit, and winter because of the cold which retains it and prevents it from exhaling, are out of order. In the south of France work can be started in March and resumed in September; but in Paris and in the rest of the kingdom, it is only in April that it can be started and the second sap is so weak that it would be a waste of time to take care of it in autumn."

This point of practice is treated entirely by Eugène Canseliet (Mutus Liber, Alchemy explained on his classic texts) and by

Atorène (the Alchemical Laboratory).

We saw it above, on the fourth plate of the Mutus Liber, then confirmed by the anonymous author of the Récréations, the harvest of the dew must be done in spring, in calm and open weather, the moon being in its ascending phase , from the first quarter to its fullest.

Of course, harvesting must begin at sunset and end in the morning. Care must be taken to only operate on uncultivated fields for fear of carrying fertilizers or pesticides, by passing a clean sheet, previously rinsed in rainwater, over alfalfa or sainfoin fields to like Canseliet.

The dew thus *"exalted"* can be more easily clarified by siphoning the supernatant liquid or by decantation, an operation necessary to eliminate the twigs and earthy particles accidentally collected.

A supply of around twenty liters will be collected and, for storage, commercial opaque plastic tanks can be used, carefully washed and rinsed with rainwater or distilled water.

Preparation of the salts

According to Eugène Canseliet, dew radically transforms the salts by making them *"philosophical"*, which will then cause the appearance of green enamel during the preparation of the repel.

The dew collected contains a salt that must be crystallized at the same time as the tartar-saltpeter supply.

You can obtain the essential equipment for this wet part as an appetizer from specialists in laboratory equipment and industrial glassware for chemistry.

Here is the procedure exposed by the fifth and sixth plates of Mutus Liber, on pages 101 and 106 respectively. It should be noted that these operations must be repeated as many times as necessary to obtain the adequate quantity of salt, which cannot be less than 1 kg in the final.

In the glass retort, preferably five liters, and two-thirds full of dew, heat over low heat to 90°C and pour approximately 300 g of tartar and as much saltpeter, gradually stirring with a glass spatula until completely dissolved.

The solubility of salts in water varies greatly depending on the presence of other salts. In chemistry, we talk about mutual solubilities, and it is quite difficult to establish these solubility curves as a function of temperature with precision. In the present case, saltpeter being the more soluble of the two salts, this will result in a reduction in the solubility of the tartar, even when hot. This will need to be taken into account when moving to the experimental stage.

We will cover the retort with its capital, with an arrangement comprising a cooled coil acting as a condenser, then we will increase the heating moderately until it boils. Slow distillation will be continued until four-fifths of the distillates have passed.

If you have placed an immersion thermometer in the retort, it will mark a temperature slightly above 110°C at boiling, at a constant atmospheric pressure of 1020 mb, and which will increase significantly until the end of the operation. The English call this known phenomenon boiling point rise, BPR, in French temperature elevation. The BPR is specific to a salt or a defined mixture of salt, and it varies slightly with the boiling temperature (and therefore pressure).

The delicate stroma collected from the retort will be placed apart in an opaque container and tightly sealed. It will be slowly digested in three or four vials for forty days. One detail, I don't believe this long digestion phase is necessary because I don't know its contribution in this salt preparation process. I think the author wanted to take us into one of the fermentation phases of the wet process.

Clearly, there are many points of practice subject to dispute. In this discipline, you should not take everything at face value, especially when you can make logical or experimental arguments.

At the end of this digestion, if you wish to carry it out, the liquid collected will be distilled again, in the four-fifths fraction as before.

The new stroma obtained must then be mixed with the stroma resulting from the first distillation.

The final operation consists of evaporating excess water from the stroma over low heat. After cooling, we will collect our crystallized salt in the manner of the bride, as we can see on the seventh plate of Mutus Liber. His august gesture represents the method of preparing real *"cream of tartar"*.

The woman is holding a vial on which we notice the four stars representing the *"harmoniac"* salt of the philosophers and we finally understand why it should not be confused with the common commercial ammonium chloride. According to followers, harmoniac salt or Ammon salt, symbolized by the ram, achieves harmony between Heaven and Earth.

Finally, an important detail that is rarely mentioned in hors d'oeuvre preparations; the enriched salt must be dried over low heat in a porcelain crucible, when damp, it would oppose the ingredient.

The distillations described by the fifth and sixth plates of Mutus Liber are just touched upon by Cyliani, in Hermès Unveiled, who presents them in the conventional form of a dream, a journey from a hot region to a cold region. These distillations are completely unknown to the anonymous author of the Hermetic Recreations.

A good tip: store your salts in well-stoppered, opaque containers, in a dry place and protected from light.

Another remark: Canseliet misled his readers by recommending that they use the two salts together.

*
* *

Plate 12 of Mutus Liber - Exaltation of dew

First Work

It is from the first work that the work of the alchemist really begins.

At this level, and if we do not then leave the framework of the dry method, the process is entirely metallurgical and the necessary equipment is reduced to an oven, a few crucibles, an ingot mold and tongs, a pyrometer.

Taking into account the products used, the greatest caution will be the rule.

Some tips for the beginner:

– all operations must be carried out under a fume hood protected by a safety glass window, or by a plexiglass panel, this will reduce damage in the event of an explosion,

– the fume hood will be suitably ventilated by an air extractor connected to a duct opening onto the roof. An electric fan of approximately 100 m3/h will do the trick, this corresponds to a standardized PVC conduit diameter of 100 mm.

An important detail, you must always leave the window ajar to allow the passage of air and avoid confinement under the fume hood.

– athanor: we can advantageously refer to the precious indications of Atorène, expert in the art, if we wish to create our own Athanor. There are many oven manufacturers on the market, and, as standard, there are gas ovens of different capacities, delivered with regulator and pyrometric probe. You can also make an electric oven yourself, as the resistance moves back and forth to eliminate the creation of electromagnetic fields. The advantage of the electric oven is that it can be equipped with a precise thermostat and incidentally a digital programmer. The temperatures that must be reached are relatively modest, they peak at 550 to 600°C. The important thing is reliability, the oven must be able to operate continuously for several weeks in a row,

– crucibles: the type, size depends on the nature and capacity of the furnace and the chemical properties of the reagents. Today we can find different types of crucibles, with or without lids, in porcelain, alundum, alumina, nickel, zirconium, graphite, silicon carbide, etc. Alundum and alumina crucibles being the most common, they are cylindrical or conical in shape, from low to high heights. For our work, we use conical crucibles of great height, with lid, for the first work and the great cooking, and always of conical shape, but of medium height for the second work.

The practice

We now have all the ingredients required for this art, as well as the equipment necessary to carry out our business.
- Our *Kohl, Ares*, finely powdered stibnite, 2 parts,
- *Aries*, iron filings, more vigorous than Ares, 1 part,
- the *Igneous Mediator*, very dry, the salt double, 1/15th of the whole.

The process described in thinly veiled terms by Flamel has since been taken up by the greatest masters, with more or less operational details.

We will also find descriptions of this process in authors such as Basile Valentin, the Cosmopolite, Philalèthe, Monte-Snyders, Henri de Lintaut, Nicolas Vallois, Cyliani and Fulcanelli. Notwithstanding, the proportions of the ingredients vary within a relatively wide range – which is, however, understandable – from 9 parts of stibnite to 4 of iron, to 8 parts to 4, or in a ratio of 2.5 to 2 to 1. These variations can come from differences in purity from one ore to another or from particular purification techniques.

Regarding this first reaction which falls within the field of chemistry, the chemist would say that, in order to establish the ideal proportions, it is necessary to take into account the purity of the two reagents and correctly measure the excess iron. Indeed, the excess of one or other of the constituents can have a fundamental impact on the final result. A lack of iron will leave traces of sulfur in the regulus, but too great an excess will lead to obtaining an ingot of antimony doped with iron, which can prove troublesome in certain applications, except, perhaps , in alchemy. Since antimony was used industrially and until the beginning of the last century, the staring – a star-shaped refill on the ingot – was considered, in commercial transactions, as the absolute index of purity of the repel.

But, contrary to popular belief, the star, ultimately, is characteristic of a regula slightly polluted by iron.

Let's dive back into elementary chemistry. In a classic formulation, for pure bodies, here is the stoichiometric reaction:

$$Sb_2S_3 + 3\ Fe \Rightarrow 3\ FeS + 2\ Sb$$

Taking into account the atomic masses of each element, the stoichiometric quantities in the same order as above are:

$$339.72\ g + 167.55\ g \Rightarrow 263.75\ g + 243.52\ g$$

Starting from chemically pure bodies, the ideal Stibine/Iron proportion can therefore be calculated:

$$339.72 / 167,55 = 2.0276\ to 1$$

Our mineral cannot be used in its raw state; it must undergo a first enrichment preparation, liquation, which allows approximately 90% purity to be obtained.

The correct ratio, taking into account the real purity of the ore previously enriched, becomes:

$$339,72 / (0,90 \times 167,55) = 2,253\ to\ 1$$

For the nitrate, decomposition takes place as follows:
$$2KNO_3 + O_2 \Rightarrow 2K_2O + 2 NO_3$$
or
$$2KNO_3 \Rightarrow 2K_2O + 2 NO_2$$

In the hypothesis where we wish to use 2 kg of sulphide containing approximately 90% stibnite after liquation, the theoretical calculations give us:

At the beginning :

Stibnite (Sb2S3) at 90% purity	2,000.0 g
Iron - ratio 2,253 : 1	887.7 g
Double salt, 1st purification, 1/15^e	192.5 g
Total	3,080.2 g

At the arrival:

Antimoiny, Sb	1,290.3 g
Iron sulfide,	1 397.4 g
Tartar	96.3 g
Potassium oxyde, K_2O	56.3 g
Antimonial gangue, gases	239.9 g

These are of course very theoretical calculations, not very analytically verifiable, for operations carried out in the laboratory, because the yields obtained are generally much lower than those in industry.

We will not be able to measure values such as the quantities of gas emitted during the reaction, the losses by volatilization of antimony in the form of volatile oxides and the mass of residual gangue. The figures above nevertheless inform us perfectly about the fundamental parameters which are the weights, and about what the ancient alchemists meant by the weight of nature and the weight of art in this first operation. They were unaware of the mechanism of stoichiometric balances, they thought that the balances were established by themselves randomly and that the weights of nature governed the balances.

It will be noted in passing that Flamel and Philalethes used much less iron than Fulcanelli who undoubtedly knew stoichiometric calculations. Starting from the clues graciously disclosed in the Philosopher's Mansions, the weights for Fulcanelli are deduced as follows:

At the beginning:
Stibine (Sb_2S_3), after liquation,	2,000 g
Iron - ratio 2:1	1,000 g
Double salt, 1st purification, 1/15th	200 g
Total	3,200 g

As we can see, Fulcanelli used a largely excess iron dosage, i.e. 12.7% more than the theoretical value. It is also possible that the starting ore, very well enriched by liquation, imposed these proportions. In any case, the purity of the ore could not be absolute. But this is perhaps an anomaly because excess iron = iron oxide in the secondary work. The deviations between the theoretical proportions of iron and those recommended by alchemists therefore appear in a relatively wide range, reaching up to 20%, with antimony sulphide in excess most of the time.

The question arises as to what importance these values have in the alchemical process. But an excess of iron oxide can hinder the progress of the finishing work.

Concerning the proportion of salts in relation to the mass of the reagents used, only Fulcanelli indicated values as modest as 1/15th of the Iron-Stibnite mixture, this proportion varies between 1/5 and 1/20th.

The rare authors who alluded to salts, such as Flamel or Tollius, mentioned much larger proportions. Tolluis, moreover, forbids the use of iron in this operation of the first work.

Some authors have used up to 50% salts compared to the other two constituents, which is enormous.

Plate 5 of Mutus Liber - The secret distillation

What could be the explanation?

Would the dosage of salt be of any relative importance?

Other authors, perhaps more informed, mention only one salt, nitre. In any case, he knew that the two salts ended up alkanizing to ultimately produce only potassium carbonate.

We can only formulate hypotheses on the proportions, but on first examination, it does not seem, beyond a minimum threshold, that the proportions of salts have a real impact on the final result, although the quality of the vitriol can feel it.

The salts react little with the other two constituents, or, failing that, with some impurities contained in the ore used, depending on its origin. Salts especially protect against oxidation and significantly reduce the melting temperature.

They will, however, have a fundamental impact on vitriolic production, which depends both on the intimate nature of the protagonists but also on the temperature and duration of the operation.

Three conditions are therefore essential for obtaining green salt:
 – the nature of the base materials and the saline mediator
 – the calcination temperature
 – the duration of the process.

But we can envisage in the bitrate molecule a substitution of the organic radical in decomposition to give either a sulfide or a potassium sulfate partly mixed with potassium carbonate and a potassium ferrite in variable proportion.

Before Fulcanelli, we find no such precise indication concerning the role of salts in the alchemical corpus. It seems that Fulcanelli, after having collected all the technical and scientific information available at his time and dealing with the methods of purifying antimony, naturally deduced that the right proportion should not be too far from 1/15th than the industry practiced for the sake of economy.

It should also be noted, and still with regard to stibnite purification techniques, that manufacturers across the Channel used a mixture of sodium carbonate and saltpeter. The proportion of sodium carbonate was around one eighth of the mixture but that of saltpeter was not clearly specified. This particularity has the merit of reminding us that potassium carbonate could be used instead of tartrate, which is chemically valid.

I will come back later to provide new clarifications on the function of salt and the secrets that no one has revealed to date.

Let us return to our process with the complete summary of the operations of the Great Work, as Bernard Husson was kind enough to leave us.

It is a gift of inestimable value, a very revealing summary of the process of the brief path, shorter but more precise than the Emerald Tablet, written by a little-known alchemist named Leona Constantia:

"To conclude, I will explain very truthfully to the kind reader, although in a different way, how he must prepare our stone:

Let him make the two warlike heroes Saturn and Mars fight (although the first is in a peaceful mood). After three or four repeated attacks, they will make peace and, as a testimony of reconciliation, will show the splendid banner, similar to a star. To these warlike heroes, now united, and somewhat tired by their ardent combat will be offered, to restore them, the water of life (which however still requires rectification) by the use of which these triumphant duelists will tie a bond of alliance forever indissoluble. As a sign of this solid and immutable union appear the two doves of Diana, holding the olive branch in their beaks. And so that this peace is announced to the whole world, a herald appears who proclaims it with a resounding voice, 7 or 9 times, throughout the universe. Now those who were contrary find themselves united; now, after many storms that have torn the rocks to pieces, after the earthquakes, after the devouring fire, there is again a whistling sound, soft and quiet. Let him who has ears to hear, hear, for I can assure that all art is contained in these few words so clear to a son of Art that there is no need to add more others."

THE PHILOSOPHER'S

Plate 6 of Mutus Liber - Second distillation

Leona Constantia is one of this small number of Adepts who have obtained the caput mortuum and evoked the sound manifestations specific to the great cooking of the dry way.

The above statement can be translated into plain English:

– Saturn represents our scaly dragon, the first matter, our finely ground sulphide from prime liquation, while Mars represents iron.

– The three or four repeated attacks correspond to the crucible tests, over open fire, of the first work.

– Rectified water of life symbolizes the saline mediator, the double or triple salt.

– Diana's doves only appear when the purification is complete and the double salt comes out as white as before use. The green coloring (the olive branches) characterizes the green lion of the first work, our mercury.

– The herald proclaiming peace with a resounding voice 7 or 9 times relating to the sonic indices of the coction which follows the Eagles or philosophical sublimations of Phylalèthe, the matter in full transformation letting out the whistles mentioned by Canseliet, at each level of increase in fire.

The devouring fire which will manifest the soft and quiet hissing relates to the end of the cooking, when the vase breaks.

The sybilline phrase *"Let him who has ears to hear, hear"* is put there, on purpose to underline, in a redundant way, the sound cues marking the coction mentioned above.

*
* *

Operations of the First Work.

At the risk of rehashing the same cautionary recommen-dations, let's recap together the usual precautions:

— you must have new, dry and very good quality crucibles,

— it is necessary to operate under a properly ventilated fume hood, in order to avoid being inconvenienced, or even poisoned, by the fumes resulting from the reaction, or in the open air,

— to heat everything, a Bunsen burner will do the trick with natural gas, or a bottle of butane as a source of energy, but it is better to use a charcoal oven,

— you must operate in small fractions of 150 to 200 g maximum and keep your cool in all circumstances, without rushing.

In a crucible containing iron filings and pulverized sulphide, all heated to around 700/800 degrees, we will pour each fraction of salt in small doses, and we will immediately put the lid back on the crucible to limit the risk of damage. oxidation.

Each time a fraction of salt is poured, there is a small detonation, it is the song of the Swan, which, according to Basil, is dying. We will see later why alchemists recommend killing the quick to resurrect the dead.

Saltpeter decomposes from around 350°C, while the reaction can reach or exceed 800°C, it reacts by detonating on contact with impurities present in the mixture and the sulfur fixed on the antimony.

— The salt must be poured in two or three times to avoid piercing the crucible.

— The required quantity having been introduced, everything is left to heat for half an hour to one hour in order to complete the reaction and homogenize the mixture. Then, we take care to invert the crucible into an iron horn or a mold made of two pieces, previously greased, then we strike it sharply to cause the separation of the regulin fraction. This operation is called Separatio.

Let us recall the recommendations of Atorène:

"For six pounds, around twenty fractions will therefore be necessary. Everyone will be kept in the oven long enough for the merger.. Each will be kept in the oven long enough for the liquation to be complete – the density of the first regula should not be less than 6.7. To do this, it is necessary to ensure that the temperature is sufficient for about an hour to keep the slag in fusion."

Further on he takes care to add:

"*...during cooking, the rest of the salt should only be added little by little, otherwise the vase will crack, especially if it is not soaked.*"

Once separated, we obtain a bright white repel and brown feces. It is necessary to repeat the purification operation two or three times so that there is no more slag on the surface, thus we will obtain at the last purification approximately 1/3 of the ore used, a proportion that is discreetly recalled. the mystical figure of Diane de Poitier (the moon of third weight).

This operation was formerly called "*The beheading of the Moor*" (the Moor is our Al Kohol, the black man).

Regarding this caput mortuum, here is Canseliet's feeling:

"*The artist, at his beginnings, would be grossly mistaken if the idea came to him that he had to reject as useless and worthless this surprising and curiously homogeneous chaos, which is also called the dead head - caput mortuum.*"

It is also with regard to this worthless residue that Fulcanelli very charitably warns the reader of the senseless appearance of alchemical operations, in the light of official science. Indeed, what experienced chemist would think of reacting the slag coming from a metallurgical refining operation with the refined metal?

This would practically be equivalent to wanting to recombine a purified body with its own impurities.

At the last reiteration we should see the perfectly drawn star appear on the regula. If it does not appear then, it will never appear. This star is caused by the formation of sinks during the slow cooling of the ingot, it is a phenomenon well known to metallurgists. The alchemists insist on the necessary appearance of this sign, a certain means of identifying the initial subject of the work. But these stigmata are not essential for the rest of the operations. In my opinion, the ancients pointed out this property to unequivocally designate the secret matter of the Great Work.

Above the star we then pick the yellowish base which will very quickly turn green if the artist has correctly applied the canonical process...

We will thus have harvested for an implementation of 2,000g of stibnite:
– regulates, approximately 1,200 to 1,300 g,
– ferric slag, approximately 1,500 g,
– vitryolic salt 200 to 300 g.

The potassium nitrate has decomposed into potassium oxide and nitrous gas which will escape during the reaction, carrying other gases along the way which will emanate a foul odor known as toxicum venenum.

At the last purification the salt will come out white.

Potassium acid tartrate transformed into potassium carbonate and sulfide, carbon dioxide and water vapor.

Potassium carbonate, sulphide and oxide; these are, without doubt, the true constituents of this mysterious Alkaest.

There is an important detail here to point out for the hermeticist: the Masters assured that the first operation consisted of reincruding the metal, which Fulcanelli contested, for whom reincruding consists of reducing the ore into subtle powder.

This is not sufficient as an explanation, Fulcanelli wanted to conceal an important detail concerning iron, which, reacting with the sulfur of stibnite, is transformed into iron sulphide and whose chemical formula, FeS, corresponds rigorously to that of natural sulphide of iron, pyrite.

The old alchemists were both very learned and clever, so the axiom Kill the quick and resurrect the dead then takes on its full meaning: in this first operation, the artist has indeed killed the quick (the stibnite), to resurrect the dead. dead, (the iron-metal killed by the fusion), by making it regain its primitive form of living sulphide (FeS).

Everyone knows that today, for the ancient masters, the refined metals resulting from human industry were considered dead, but the ores from which they came and which had not yet undergone the action of fire, seemed alive to them.

I must make a warning about the manipulations described above. However, they correspond point by point to the process described by Canseliet in "*Alchemy explained*" then by Atorène in the "*Alchemical Laboratory*", but unfortunately, they do not lead to the true and ancient fountain.

In reality there is a big difference between the process described by Fulcanelli and that of his student.
This difference lies in the application of salts. The curious student will be able to profitably compare the texts of the Master and his disciple.

This constitutes an unprecedented revelation of which I have reserved the first for you.

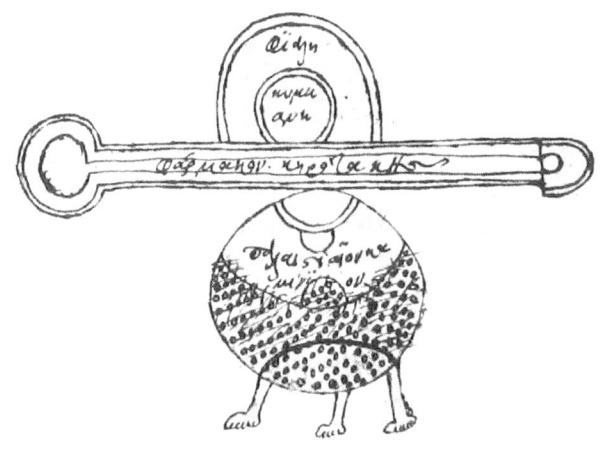

Second Work

After completing the first work, alchemy takes a route that resolutely departs from the paths of modern chemistry or metallurgy.

Nothing, after this preliminary phase, seems coherent or rational.

This is also where the dry-wet routes separate, and at this stage several variants are proposed by some enthusiasts.

Both routes, dry and wet, start the same way.

For Fulcanelli, the processes differentiate themselves after the first work. He specifies, in his commentary relating to Flamel's initiatory journey to Compostela, that the two paths begin with the same process until the sublimations at the end of which they separate.

The aim of the operations of the second work is to obtain the philosophical or double mercury, the rebis:

– in the case of the dry method, the alchemist will endeavor to subject the compound to a series of crucible operations, called Sublimations or Eagles.

– in that of the second way, the long or wet way, it involves subjecting the alchemical earth to a slow wet dissolution at ambient or moderate temperature, alternated by phases of drying then new dissolutions and so on until to obtain sulfur. The double difficulty of this route lies in the identification of the secret solvent and in the slowness of the imbibition operations.

That's not all, there are also variations for these two routes, they both involve the use of ordinary mercury (Hg) and natural gold.

Fulcanelli immediately categorically rejected the hypothesis of this third way in his first book, The Mystery of the Cathedrals. Notwithstanding, Flamel, Philalhète, Urbiger, Le Vallois, Kerdanec de Pornic implement the exalted hydrargyrum well.

It must be remembered here that alchemists have always maintained confusion between dry and wet methods, between common mercury and our mercury, with the aim of confusing the layman.

Even Eugène Canseliet allowed himself to be fooled into thinking he could achieve the animation of gold with mercury, through slow exposure to the heat of the amalgam placed in a glass balloon. In his Alchemy explained on his classic texts, he explains in detail the process he followed in 1931.

"We first proceeded to modify the mercury, by the ordinary method of distillation, in sandstone retorts or retorts, which are completely unobtainable today. As for the gold, so that we were certain of its total purity, we subjected it to the test of lead and the dish, and we then made a trichloride from it, by dissolving it, in fragments, in the 'aqua regia. We evaporated gently in a sand bath, then in a water bath, we freed our crystals of superfluous hydrate. The metal precipitated from its solution, that is to say reduced from its chloride, appears as a fine powder which, when well washed, is very suitable for amalgamation. The mercury of the seeded kind, that is to say by amalgamation, is introduced into a matra with an ovoid or spherical body, made of glass of good thickness, immediately closed with a solid lut, otherwise with the blowtorch. Then the cooking of the amalgam is regulated and maintained without failure, for many months and always below the boiling temperature."

This type of operation can only result in the manufacture of an amalgam, nothing other than a simple inert alloy, incapable of causing the slightest reaction within the harshly chemical mixture. The alchemists of the Middle Ages knew that *"bodies have no action on bodies, and that only spirits have an action on bodies"*. Was the disciple of the great Master, a specialist in the dry way, unaware of the true wet process, that of Cyliani or that of Huginus at Barmâ?

Fulcanelli assures that, for the long route, it is necessary to start from natural gold dissolved in philosophical mercury (it is therefore not the common mercury used by Canseliet). For Cyliani and the anonymous author of Récréations Hermétiques, the common core of the two paths ends when the star or regula is obtained at the first work, that is to say just before starting the sublimations, and therefore a step before that indicated by Fulcanelli.

The path described by Cyliani and which leads to sublimations is indeed a wet path, it consists of treating a powdery earth with a liquid "spirit", it is a sort of liquid-solid extraction associated with reactions which occur at temperature ambient.

However, the message of Flamel or Philalhète seems clear: the process that they both explored consists of modifying flowing mercury, by subjecting it to complex and repeated distillations in amalgamation with philosophical mercury. Their path is therefore completely different from that explored by Cyliani, Albert Poisson or Canseliet among others.

Here are the details of the operations described by Flamel in his Breviary:

"So it is male and female and our hermaphrodite which is mercury and is this work in image on the seventh leaf and bonus of the Jew Abraham which is namely two snakes about a golden rod, as you will see in This is a book that I made myself to my liking as best as I could imagine it for clairvoyance and a philosophical document. So be advised to use good supplies and ammunition because it is a job to have a lot of them, that is to say 12 or 13 pounds or even more depending on what you want to do in many operations.

"So you will marry the young Mercury, that is to say quicksilver, with him who is saturnial philosophical mercury, so that through him you can avoid and strengthen the said quicksilver current by 7 or even 10 to 11 times with the aforesaid agent who is called a key or sword of sharpened steel, so that it cuts, incises and penetrates the body of metals, and when done you will have such mastery, then you will have double and triple water painted in the image of the rose bush from the book of Jewish Abraham which comes out from the foot of an oak, namely from our saturnia which is the equal key, and will precipitate itself into abysses as the above-mentioned Jew says, that is to say in our receptory which is arranged at the neck of the retort where the above-mentioned mercury will be thrown, doubled by the art and device of a proportionate and suitable fire.

"But here lies an anxious thorn with even impossible to work if God

does not reveal the above-mentioned secret or the master does not open it, because Mercury does not marry with Saturnia regal without a thing that is hidden in a right device to be examined as is done and plowed, because if you do not know the machine how the said valor is done and peace with the aforementioned quicksilver, you will find nothing of value. So dear and beloved nephew, I do not want to hide anything from you but tell you everything without keeping anything and show you how I must rightly advise the fact and the point that is profession in this philosophical mastery, so tell you that without sun or no one will benefit you from the said work. You will therefore make him eat our old man or voracious wolf gold or silver in weight and measure as you say. Give full attention to what I say so that you do not err and fail as I did in this task. So how can we feed our old dragon gold? Advise right machine in natural reason, because if you give a little gold to the Saturnia fuse, it is very much apparent, but quick silver will not come to life this is an incongruous thing which is not even profitable, and I have greatly labored in sadness before finding the right machine to do this. So if he has a lot of gold to devour, he will not be so open and ready, but will then take it quickly and they will both marry each other in dough. Do as you saw me do. Note that you have to work in everything according to the weight that tells you because without that you will not plow for your benefit but to your detriment, record this, here is the machine found. Therefore seal the said secret because it is everything and never write it on paper or anything else that can be seen written because we would cause damage to worldly universality. But I give it to you under the seal of the secret of the conscience in love that I have towards you. Advise to take so this one fuses, throw in then throw this one into a marble, crushed into powder with 12 quicksilver. Make these take like butter or cheese by pounding and stirring here and there when and when the other and washing with clear common water until the water comes out clear, and the mass appears clear and white (so you will do on moon fuse) then is made conjunction of it with the solar regal Saturnia. When is now as well as butter, take the trap that you dry gently with canvas or fine cloth many machines. Here is our lead and our mass of nonvulgal Gold and Moon as well as philosophers, so put it in a good mix-

ture of earth in a crucible, much better steel, then in a furnace, and set fire, going little by little. Arrange a receptacle for the retort as is profession two hours, and after vigorously your fire as long as the mercury is in the receptory above said, and is this one mercury the water of the flowering rosebush, even the blood of the innocent slain in the book of Jewish Abraham, therefore being this water of Gold and Silver philosopher. So believe that this mercury ate a small part of the body of the King and that it will already have much greater strength to dissolve the other one after which it will be much more covered with the body of Saturn."

"So you move up a degree or rung on the ladder of art. There take the feces of the retort, fuse them in a crucible over high heat and bring out all the Saturnian smoke and when the molten Gold is clean, infuse it like the prime times two of Saturnia. Therefore is the sun IX infused in the said feces much more apparent than the first time. And as mercury is already more sour than it was before, he will already have much more hidden strength and vigor to scrutinize and so to speak eat it again and to fill his belly little by little. So advise dear nephew the degrees of the engine of Nature and Reason, in order to ascend by steps to the highest part of Philosophy which is throughout the course of Nature and which you would never have found if you did not give up this mastery. So bless the Lord for what I have given me to want towards you, for without this we would have worked nothing as some do with the loss of many pecunes, infinite pains and labors, anxious vigils and costly cures. So do the same as the first times, combine with the mercury released above and strong in degree by grinding and pounding so that all the blackness comes out, dry as you said. Put everything in the above-mentioned retort and do no more or less everything as you have just plowed for two hours at a low and suitable heat, then strong and good to push and let the mercury flow into the receptacle, and you will still have a lot of mercury further accused, and you will be ascended to the second level of the philosophical ladder. Do and work again as you have just worked by throwing the Saturnian son into the appropriate weight, that is to say little by little and working with the machine neither more nor less as you worked at the beginning, as long as you are at the tenth degree of the ladder, and then

you rest and is already said igneous mercury acuity, fully engrossed, big with male sulfur and vigor of astral salt that is in the deepest caves and viscera of gold and of our Saturnian dragon, and believe that you are writing something that no Philosopher has ever said or written."

We will finally have understood that there is a third way, the way mentioned by Flamel which is a variant of the primitive way. It is accomplished with exalted mercury (Hg), conventionally called our mercury, acting as a *"fugitive servant"*, to extract sulfur from gold (Au). This path is different from the canonical path which uses the Alkaest solvent to extract philosophical sulfur from alchemical gold.

This is also Fulcanelli's feeling: *"the dissolution of alchemical gold by the Alkaest Solvent characterizes the first way; that of common gold by our mercury indicates the second. Through this we create animated mercury."*

In the ancient way natural gold is excluded as well as flowing mercury because it has no effect on iron or its compounds.

Purification

Arriving at this stage of the process, most authors remain silent and resume their comments at a more advanced level, omitting an entire regime absolutely essential for the successful realization of the Stone. This pitfall is pointed out by Fulcanelli who, after Limojon de Saint Didier, attacks Philalèthe, accused of having adulterated the order of operations and falsified the reagents by introducing hydrargyrum into his process. But, if the reader has properly understood Flamel's operations in his Breviary, he will be able to see that Philalethes is only following the path described by his illustrious predecessor and there is no need to blame him.
Limonjon de Saint Didier exposes a great secret, that of the two

mercuries: *"The woman who is proper to the stone and who must be united to it is this fountain of living water whose source, entirely celestial, which has its center particularly in the sun and in the moon, produces this clear and precious stream of the wise , which flows into the sea of philosophers, which surrounds everyone. It is not without foundation that this divine fountain is called by this author the woman of the stone; some have represented her in the form of a celestial nymph; some others give it the name of the chaste Diana, whose purity and virginity is not sullied by the spiritual bond which unites her to the stone. In a word, this magnetic conjunction is the magical marriage of heaven and earth, of which some philosophers have spoken; so that the second source of the physical tincture, which works such great wonders, takes birth from this very mysterious marital union."*

The advice of the Adept relates to philosophical sublimations, in other words the Eagles. Canseliet explains that these operations were veiled by Flamel under the allegory of the massacre of innocents. This allegory, he explains, relates to the splitting of materials to facilitate sublimation.

In short, the whole secret of the Eagles consists in reacting metallic antimony, and potassium nitrate.

We collected from the first caput:
– regulus,
– ferric slag, Fulcanelli's Adamic earth,
– vitryolic salt.

There are size anomalies if we refer to the proportions required for the finishing work, we must now reveal to you the secret of the weights.

The first work, for 2 kg of pre-treated stibnite, delivers approximately 1.5 kg of ferric slag which will then require 3 kg of repel and 675 g of salt to undertake the Eagles.

However, at the end of the first work, the reaction theoretically gives us 1.3 kg of regulus.

You must therefore take all precautions from the start and stock up on a good supply of common mercury as well as salt if you do not want to be stopped at the second job.

You need "*plenty of distilled spirit water*", according to Arnaud de Villeneuve and the Nymph of Cyliani. The weight of water must be plural, according to another alchemist.

The said water is the potassium nitrate.

The artist will therefore be advised to use at least 6 to 7 kilograms of ore, the excess of which he will reserve for the operations of the second and third works. It will also make a good supply of salt.

The secret of the purifidation was revealed in a manuscript by an unknown follower, it involves the purification of the regulus or otherwise the very secret sublimation which is done in 3 operations.

These operations are always carried out in the crucible with all the usual precautions. It is in fact the repetition of a single operation.

We melt 1500 g of regulus in a crucible until red/white, provide around twenty packets of 10 g of dry potassium nitrate wrapped in paper. We therefore throw the reguus in fractions into the white-hot crucible. We wait until the previous portion is well melted before adding a new one.

At the last, we violently push the fire after having covered the vessel, so as to bring the mass to boiling point.

While the fire is burning, we prepare a large basin of cold water, a sulfured iron mold and we cover our eyes with dark glasses and our hands with wet woolen gloves.

When the temperature is reached, which we recognize by the torrent of white and poisonous smoke escaping from the crucible, we grab the lid of the crucible with our left hand and quickly throw in a packet of saltpeter. of approximately 10 grams and cover the crucible, holding the lid with the pliers.

Buzzing sounds are heard, soon followed by internal explosions, a dull noise, then clear and rapid.

At the first purification we will use 7 packets,

While the fire is burning, we prepare a large basin of cold water, a sulfured iron mold and we cover our eyes with dark glasses and our hands with wet woolen gloves.

When the total dose of niter is exhausted, the fire is left to act for a while longer through digestion, then the metal is poured into the previously sulfurized mold. After cooling, an ingot, brown on the outside, is extracted from the mold, a sort of shell formed by the union of saltpeter and arsenic;

- the *Dog of Corascene* by Philalèthe.

This gangue is very tenuous and very caustic. To avoid as much as possible the burns it causes, it is broken with a hammer in cold water. The cap of the ingot, less tenacious, separates in one piece.

It is a plug of black earth, and if the operation is successful, the ingot already bears at its top the first lineaments of the Star of the Magi.

The black hat should be thrown away.

We then undertake the second purification with 6 packets of salt, then the last with 5 packets.

Above all, be careful not to throw all the salt at once as this would cause an explosion. It would be preferable to operate in the open air to avoid breathing toxic gases. Once the last pour has been made, the ingot is struck sharply with the hammer to reveal the famous Star of the Sages, which was the purity index of antimony in the 19th century.

It is a shrinkage which appears when the metal cools, and resulting from contraction. You must collect all the soil and reserve it in an earthen pot blocked with emery.

*
* *

Eagles or Sublimations

The eagles are the repetition of the purification operations and are intended to obtain the red sulfur. It takes 7 to nine according to the great masters to succeed in the Great Work.

The purpose of the sublimations is to obtain the enigmatic fish that the alchemist must cook in his own water.

The product from sublimation is presented in the appearance of a purple texture button and a mass that a dozen grams and that we compare a few times to the return button during the purification of gold by coupal with lead.

The third board of the Mutus Liber is the richest in hermetic symbols, and shows several sinners busy attacking fish, and we notice on the lower right part of the engraving, a woman holding a line and meaning only one fish is required by brandishing her index finger, while another woman hands a fishing net. For whom is customary of symbolic representation, the net is not by chance on the engraving! Indeed when the operation ends in the ideal conditions, the surface of the bath after crystalization offers the appearance of a net with intertwined meshes. The hermetic fish is trapped in solidified mass and it must be released.

Fulcanelli compares this *"fish"* with the bean or the swimmer (the little Jesus) of the Galettes of the kings whose tradition dates back to the ancient Roman Empire, in The Mystery of Cathedrals: Fulcanelli develops an important secret: *"Our pancake is signed as the material itself contains in its dough the small child popularly emitted bather. It is the child-jesus carried by Offerus, the erviteur or the traveler; is gold in his bath, the swimmer is the bean, the hoof, the cradle or the cross of honor, and' is also the fish "which swims in our Hilosophical Sea", according to the very expression of Cosmopolite.*

Note that, in the Byzantine basilicas, Christ was said to be represented as the sirens, with a tail of fish. We thus see him figured on a marquee of the Aint-Brice church, in Saint-Brisson-sur-Loire (Loiret). The fish is hieroglyph of the stone of philosophers in its first state, because stone, like fish, is born in water and it in water. Among the paintings of the alchemical stove executed by P.H. Pfau, we see a fisherman with a line coming out of the water and a beautiful fish.

Other allegories recommend grasping it using a loose net or rets, which is an exact image of the meshes, made up of crossed wires, delicate on our Epiphany pancakes."

First Eagle

To do this, you must use the expelled product, weigh it carefully, and substitute the potassium nitrate of which you will make 7 packets.

We start the first eagle by melting the starry regulus into small pieces in a crucible. The fire must be lively to fully open and move matter in all its atoms. We then pour the weight of nitre , divided into two packs, but at a certain interval. We let it act by keeping the vase well covered, because then the eagle begins to attack the lion.

The material blackens in the crucible little by little, and swells like dough, enclosing the leaven, and a crust forms on the bath. When the compound has regained its calm, the reaction is complete, the material is poured into a sulfurized ingot mold and allowed to cool.

The cooled base has a thin pellet of modified niter at its top. It must be preserved carefully, because it is the aurific oil which will be used for the second eagle.

Second Eagle.

We take the pellet which tops the cooled base of the regulus, which pellet is our aurific oil, so named because it is impregnated with the interior gold. It always comes to the top after cooling. It is put into powder, or into small pieces; we put the crushed regulus (the base) in the melting crucible and dissolve it. When the fusion is complete, we gradually pour in the fragments of our aurific oil tablet,We then take the second dose of saltpeter. We make two packets which we throw into the molten material at a certain interval. The niter is then left to act for some time and, once the reaction is complete, the material is poured into the sulfur mold and allowed to cool.

Third Eagle to Seventh Eagle

We take the pellet which tops the cooled base. We put the molten base back in small pieces, the fusion being perfect, we take the dose n° 3 of nitre of which we make two packets, and we throw them away one after the other,It seems, from reading certain authors, that the use of saltpeter is done in the same dose in all eagles at a rate of 10 to 1.The potassium nitrate is given time to act and, once its action is complete, its fusion is poured into the sulfur mold. and so on until the seventh eagle.

*
* *

Coction - Multiplication - Projection

The great coction is the most secret operation in alchemy, and yet the great follower Fulcanelli expressed himself in a prolific way with such a skill that it is practically impossible to leave the labirinth in which it leads us.

The large coction is made in a crucible on bare fire for a period of 4 to 7 days, it consists in cooking the fish from sublimations in its own earth according to the saying of the alchemists.

According to Fulcanelli *"In the alchemical field, the Greek cross and the Saint-André cross have some meanings that the artist must know. These graphic symbols, reproduced on a large number of manuscripts, and which make, in some prints, the The object of a special nomenclature, represent, among the Greeks and their successors of the Middle Ages, the crucible of fusion, that the potters always marked a small cross (crucibulum), index of good manufacture and proven solidity. Greeks also used a similar sign to designate an earthcut. Besides, the word Matras, used in the same direction in the 13th century, comes from the Greek Metra, a term also used by the blowers and applied to the secret vase used for the maturation of the compound. century, gives a figure of this spherical utensil, tubelé laterally, and which he also calls matrix. The X also translates the ammonia salt of wise, or ammon salt (ammoniacos), that is to say ram, which we once wrote with more truth Harmoniac, because it realizes harmony (Armonia), assembly, the agreement of water and fire, that he is the mediator par excellence between heaven and earth, mind and body, volatile and fixes it. It is still the sign, without any other qualification, the seal which reveals to man, by certain superficial lineaments, the intrinsic virtues of the premium philosopher's substance. Finally, the X is the Greek hieroglyph of glass, pure matter between all, assure us the mas-*

ters of art, and that which the most approaching perfection."

At the end of the great coction, if all the operations were carried out properly, the crucible split and finally reveals the philosopal stone in its rough gangue.

Fulcanelli in the philosopher's residences *"Indeed, when the artist, witness to the combat that the Rémora and the Salamander engage in, steals the igneous monster, defeated, his two eyes, he must then apply to bring them together in one. This mysterious operation, easy however for those who know how to use the Salamander corpse, provides a small mass quite similar to the oak glans, sometimes to the chestnut, depending on whether it is more or less dressed in the rough gangue that it does not show Never entirely released."* Before carrying out the projection, it is necessary to restore the stone the metal characteristics that the coction made it lose.

*
* *

Third board of the Mutus Liber - Symbolic fishing

Practical considerations

By studying the alchemical literature abundantly published today, we are struck by the multiple facets presented by this old and noble discipline.

I would like to once again emphasize the overrated reputation of certain authors, of whom we can be assured from reading the first pages that their supposed rise to fame is doubtful. This assertion is not gratuitous, it simply arises from a critical examination of the content of these numerous works devoid of any moellific substance which an objective analysis quickly allows us to exclude from the hermetic corpus.

Good authors are recognizable by the operational details they indicate, the materials used and the way in which the process of the Great Work is presented.

However, extensive experience, coupled with excellent knowledge of the corpus, is required to form a real opinion. Fortunately, it is not necessary to have an imposing library to complete the study in which one finds oneself engaged. On the contrary, sometimes it is better not to disperse, the abundance of references increases the difficulties due to the proliferation of non-convergent means of exposure.

The alchemical processes, as developed and commented on in this work, do not come from an empirical, rapid, superficial or gratuitous extrapolation, but from a long and patient study of the best authors as well as the analysis of the different points of doctrine. , of their confrontations with experience.

The study of alchemy requires, to be properly undertaken, either a good master, a real initiator according to tradition, or the continuous and repeated reading of as many alchemical works as possible, reinforced by experimentation in the furnace. Accomplishing self-initiation is within the reach of any stubborn and persevering researcher, as ultimately many masters have achieved over the last ten centuries.

Alchemy, the esoteric Science par excellence, is a universe in which humility is the most fundamental of the qualities required to complete one's initiation.

In alchemy, everything will remain vague, uncertain, for the artist working in the furnace, as long as the ultimate transformation, that of base metal into gold – undeniable proof of success – has not been accomplished. We must be wary of all authors, without exception. The tracks are blurred beyond reason. In addition, to the intentional cabalistic encryption was added the influence of certain irrational beliefs, which I would describe as heretical, of astrological origins, and the consequences of which can be seen in many authors.

Indeed, we must question the validity of these concepts borrowed from astrology by comparing them with the axioms well anchored in the Hermetic tradition.

To illustrate this question, we can first mention the usefulness of dew in the alchemical process as defined by certain authors from the end of the 14th century and although no allusion is made to it by their predecessors.

Then, we will question the recommendations relating to the favorable season for the work, the lunar phases or the orientation of the equipment etc., knowing that one of the most important axioms specifies that the work can be done in any place and at any time.

The above is, of course, in perfect contradiction with the processes exposed by the Cosmopolite, the Mutus Liber, by Limojon de Saint Didier, by Gobineau de Monluisant, by the anonymous author of the Récréations Hermétiques and finally by Cyliani. Regarding Fulcanelli, he remained rather laconic, the Dew of May, according to him, is easily extracted from a vile and abject body, the subject of philosophy, he nevertheless indicates that it is necessary to use a second salt, extracted from the dew of May, to perfect the operations of the first work. Understand who can!

Finally, it seems necessary to mention the anomalies and contradictions relating to the processes described by different renowned authors.

Thus, we can count three different processes capable of leading to the Stone:

– the dry process, the short way, followed by Basil Valentin, Fulcanelli and his disciple,

– the half-dry, half-wet process, followed by Huginus in Barmà, Urbiger, Flamel, Philalèthe, Monte-Snyder, among others,

– the wet process or the long way, followed by Cyliani and the anonymous author of Récréations Hermétiques.

The first process seems to be the ancient path, followed by Mary the Prophetess and Artéphuis, except for the entire hors d'oeuvre part where the dew seems absent.

I will add the following few comments:

First of all, regarding the sublimation phase, the risk of "burning the flowers" is clearly the most frequent at this stage of the process, the collected return button often presents blisters. The alchemists therefore had a secret way to limit the liquefaction temperature of the compost during the second work. They would introduce another element not listed or mentioned among the most prolix authors, or perhaps they were unaware that this body was present as an impurity in the elements they used? This passing remark could also explain the repeated failures of Canseliet, the only one, along with Fulcanelli, in the last century, capable of mastering the entire process of the short route.

Fulcanelli, an expert in quality, was certainly aware of the pitfall pointed out by rare authors. Concerning this mysterious body, it seems that there are clues here and there, first in De Pontanus, then Limonjon de Saint Didier and finally Cyliani, which support my deductions.

I think I have several ideas to submit to experimentation, before communicating my feelings on this imperfectly explored subject.

The second process, which I intentionally named "*mi-dry*", is in reality an adaptation of the ancient path where natural gold is substituted for philosophical gold. The essential characteristic of this path concerns the circumvention of the difficulty of correctly performing the Eagles and the impossible skill it requires.

It would seem that thanks to the thermal balances modified by the presence of mercury (Hg), the extraction of rebis is greatly facilitated, at a much lower temperature than that of the traditional compound. In addition, a simple distillation of the compound, a mercuric amalgam, allows both to eliminate unwanted mercury, and to sublimate the active part necessary for the realization of the rebis of the concoction. Mercury has the double advantage of being inert under the reaction conditions, and of moderating the transformation temperature of the compost, while maintaining a constant temperature of 360°C until the distillation of the last atom of mercury. This process also offers the advantage of only requiring a Pyrex glass flask, fitted with a capital and a condenser. This variant, strongly criticized by Fulcanelli in the Mystery of the Cathedrals, certainly makes it possible to avoid *"burning the flowers"*.

The third process, that of the long and wet route, poses a real enigma. We saw above that Canseliet, after Albert Poisson, had allowed himself to be misled by the fallacious process (he himself says it) which involves a circulation of mercury (Hg) and gold (Au). in the form of fulminate. In fact, what he believed to be the wet process does not correspond to the path described by Cyliani, it is in reality a very bad interpretation of the variant of the "mi-dry" process described above.

The wet and watery way exposed by Cyliani begins like that of Fulcanelli, with the same hors d'oeuvre preparations including the calcination of the first work. It was certainly at the end of the operations of the first work that Cyliani left the dry route to choose the wet route.

At the start of the imbibitions, he will use a liquid element which is not flowing mercury (Hg), to carry out the long sequence of dissolutions. However, he will use the flowing mercury later, in a variant already reported for the extraction of sulfur from gold.

Yes, but then, where does this liquid element used to make the imbibitions come from?

Several hypotheses can be considered:

– that of the irruption of a new character on the stage of the alchemical theater, the Mercurial Fountain, whose source is more celestial than terrestrial, according to the words of Limojon de Saint Didier. This hypothesis is mentioned by Bernard Husson,

– or the Philosophical Water of Fulcanelli, that which must be revealed in a dream, in other words...

– that of the water collected by failure of ferric slag,

– that of dew (see Mutus Liber) or its distillation (astral spirit),

– finally that of the distillation of the salts of the work to produce the stinking menstrual, strong water.

In this last hypothesis, saltpeter, by decomposing while hot, releases the NO_3 fraction which can simply be absorbed in a stream of cold water to make nitric acid. (Strong water? Stinky menstrual?)

The NO_3 radical produced during the decomposition of saltpeter would intervene directly, in the dry process, at temperatures close to 500°C.

So why not consider using strong water (nitric acid) to carry out the reactions at moderate temperature using the wet method?

Cyliani describes a process requiring only beakers in which the alchemical material is divided into small fractions, then sprinkled with a liquid, at room temperature – it is important to emphasize this. In this process, we moisten the fixed, that is to say our Adamic earth, the slag coming from the star compost, with the volatile (nitric acid ? which is indeed volatile). This operation does not require closing the vase, on the contrary, the vase is opened to allow the evaporation and drying of the earth, then the new sprinkling of liquid and so on. This path is marked by a variation in colors reported by most authors.

From a strictly scientific point of view, modern chemistry is able to explain the acceleration of the alchemical process of the short path. Indeed, I have already indicated this, admitting that a process of a chemical nature does indeed intervene, even in a secondary way, in the Great Work, such as for example the action of the NO_3 radical on one of the compounds, it is logical to think that the involvement of this same radical at room temperature will cause the same type of chemical reaction, but with significantly slower reaction rates.

Kinetic laws predict a significant acceleration of most chemical processes, practically doubling the reaction rate at each 10 °C step for first-order reactions, as is the case here.

So imagine the time saving obtained when the reaction temperature increases from 20°C to 530°C.

The above therefore accredits the existence of the two paths.

The continuation of the operations of the wet method is amply described in Hermès Dévoilé, as well as in the Hermetic Recreations. The operations are too long to be summarized here.

If Cyliani used flowing mercury and gold, at one point, it was to facilitate the production of philosophical mercury by extracting sulfur from gold, just like the use of the same mercury in the variant of the dry way. This is all very consistent. The only handicap of the wet method is therefore due to the slowness of the reactions and the alternating imbibition-drying process. But this route is much safer than the dry route. The total duration of wet process operations ranges between 6 and 24 months according to the authors.

Another point of the wet route deserves to be mentioned, it concerns the weights. Here, no scales, the imbibitions are done by assessing consistency. Weight of art, weight of nature?

*
* *

Transmutations

There are many testimonies of successful transmutations. Entire catalogs have been written on transmutations carried out over the centuries, the latest being that written in 1974 by Bernard Husson, "Alchemical Transmutations". The learned Hermeticist also mentions the coins immediately beaten with the gold or silver of successful transmutations, in front of witnesses. Some of these coins are kept in museums.

Invariably, the Philosopher's Stone described by Adepts or by witnesses who have witnessed transmutations, always has the same properties.

Red in color tending to garnet, resembling crushed glass, very high density, it reacts effervescently with molten metals. Following an alchemical transmutation, once the casting has been carried out, we obtain a gold whose most surprising property is that it can further ennoble a certain quantity of base metal.

We also know that the synthetic gold thus obtained is of incredible purity. The goldsmiths were aware of this, to the point that they knew how to recognize gold obtained by transmutation only by the title.

In the anomalies observed during these experiments, it is sometimes necessary to note unexplained mass contributions. The gold ingot obtained by transmutation, when weighed, denoted a significant increase in weight compared to the quantity of metal initially used. In other, less common cases, there is, on the contrary, weight loss.

Detractors have always believed they saw it as the result of manipulation or cheating. They were at their expense and this for reasons known for a long time:

– First, the falsification of a metal like iron, by depositing copper for example, by displacement of equilibrium, has been known for a long time, and the creation of an alloy resembling gold is a matter of ridiculous hypotheses , is to cast doubt on the competence of the jewelers or testers of the time.

Since Archimedes, we have been able to tell the difference between gold and any alloy, using the double weighing system. Since antiquity, we have also known how to extract gold from an alloy by cupellation, possibly with prior refining with antimony.

The touchstone allowed relatively easy and rapid dosing. As a last resort, acid or aqua regia attacks could make the difference.

– Secondly, during many transmutations reported, it was not the alchemist himself who carried out the demonstration, but a witness, a notable or a goldsmith who operated in a completely neutral and impartial manner.

– Thirdly, most of the time, the alchemical gold remained the property of the witnesses of the demonstrations, some of whom refused to part with it even at a high price, which proves if necessary, the total disinterestedness of the Adept who offered these demonstrations .

There are also several non-alchemical transmutation processes that I have not yet discussed in detail, these are the so-called "*particular*" processes.

Fulcanelli refers to it on numerous occasions in his trilogy. The most profitable of these individuals is, it seems, that of Basile Valentin, a variant of which was exploited by a Tunisian alchemist. This process was reported by Saint Vincent de Paul upon his return from his captivity in Tunisia.

Basile Valentin's process starts from silver which has probably undergone a pre-conditioning treatment, associated with the alchemical sulfur of iron and copper. The following recipe is a variation of the two-cement gold described by Fulcanelli.

The great insider gives an easy-to-implement recipe.

"The moon itself has a fixed mercury by which it sustains the violence of fire longer than other imperfect metals; and the victory she wins clearly shows how fixed she is, given that the ravishing Saturn can take nothing away or diminish her. The lascivious Venus is well colored, and her whole body is almost only dye and color similar to

that of the Sun, which, because of its abundance, tends greatly towards red. But as much as his body is leprous and sick, the fixed tincture cannot make its home there, and the body flying away, the tincture must necessarily follow, because the latter perishing, the soul cannot remain, its home being consumed by fire, not appearing and leaving him no seat and refuge, which on the contrary being accompanied remains with a fixed body. The fixed salt provides the warrior Mars with a hard, very solid and robust body, from which comes his magnanimity and great courage. This is why it is very difficult to overcome this valiant captain, because his body is so hard that it is difficult to injure him. But if one mixes his strength and hardness with the constancy of the Moon and the beauty of Venus, and grants them by spiritual means, he will be able to bring about not so much mischief but a sweet harmony, through it."

In the process described by the most famous adept of the Middle Ages we see that iron seems to play a preponderant role in the transmutation processes. This clearly confirms the importance of this metal in the Great Work as in most particular processes, which makes Limojon de Saint-Didier say to Pyrophiles that:

"The success that some artists have had in certain particular processes comes only from the stone."

Fulcanelli, in the first volume of his Philosophical Mansions gives us descriptions of particular processes using silver and iron, because he says, iron is the metal which has the most affinity with gold and we can easily extract from iron a mercury and a sulfur which can be used in so-called special processes to generate or increase gold. But copper can also be suitable.

In the process described by Saint Vincent de Paul, upon his return from captivity in Tunisia, the silver, in sheet form, is interposed between sheets of gold previously enriched with sulfur extracted from iron, in the presence of a flow, probably containing saltpeter mixed with crushed brick. After spending 24 to 48 hours in an oven set at a temperature lower than the melting temperature of the alloy, it is then heated until melted and a sparkling gold ingot is poured. Exalted, coral gold, according to Fulcanelli, is always capable of transmuting a certain quantity of silver by inquartation. Although Fulcanelli maintained that the particular processes have no connection with alchemy, it must nevertheless be recognized that all archchemical or spagyrist transmutations must be placed in the same category as those of alchemy because they belong to the same order. physical phenomenon and it could not be otherwise.

An important detail: nitrogen seems to play a major role in transmutations, just as in the development of the Philosopher's Stone. Is this just a coincidence?

In the register of transmutations based on so-called particular processes, it is interesting to note the companies carried out in parallel, at the end of the 19th century, by two hyperchemists. These are the experiences of Tiffereau and Doctor Emmens, recounted in a curious work attributed to J. Marcus de Veze, Alchemical Gold1. These two researchers, working separately, and employing two processes, apparently distinct, but which had in common the starting metal, silver, and a chemical reaction with nitric acid. They succeeded in weight transmutations from silver to gold. These experiments have been attested by several trustworthy witnesses, moreover Doctor Emmens managed to sell several ingots of synthetic gold to the Government of the United States of America.

Tiffereau, by subjecting silver alloyed with copper to solar radiation, then under certain conditions, to nitric acid, obtained gold, following numerous manipulations carried out by alternating nitric attack and evaporation, a process resembling at the Cyliani wet way. He successfully repeated this experiment twice in Guagalajara, Mexico, and transformed all the alloy used into gold.

On his return to France, with his successive successes in Latin America, he tried experiments with money again but without obtaining the slightest success.

Doctor Emmens, for his part, had developed a mixed, mechanical and chemical process. Starting with Mexican silver dollar coins, he subjected them to mechanical compression using a DIY press weighing several tons. He then melted the silver and subjected the metal to a chemical attack with nitric acid. After refining, Emmens recovered around 30% pure gold, coming from the transmutation of silver.

There is a lot of analogy between the processes described above and those of the individuals mentioned by Basile Valentin and Fulcanelli, as well as the wet process. Tiffereau and Emmens used Mexican silver which they subjected, although in a different way, to the action of nitric acid. Tiffereau, who did not achieve the slightest success in France, attributed his failures to the hostile climate of the metropolis, less sunny than in Mexico.

In the experiments carried out by Tiffereau, we note the use of an alloy of copper and silver subjected to the energetic and repeated action of nitric acid. He was on the verge of rediscovering the particular nature of Basile Valentin.

Finally, I would like to point out an important detail, the use of nitric acid which comes up in many so-called "special" processes, so remember the few thoughts raised above on the wet alchemical process.

It is therefore no longer a coincidence that we find nitrogen once again, in a transmutation process,

*
* *

BNP-Paribas Rue Bergère Paris - The Portal of Hermès Trismégiste.

The Portal of Philosophy

Paris is certainly among the European cities best provided for works of art, and more particularly with regard to monuments erected to the glory of Hermés.

On the portal of the old Comptoir d'Affompte de Paris, at number 14 rue Bergère, now sheltering the offices of the BNP-Paribas, in the heart of the ninth arrondissement, the artist sculptor Aimé Millet signed a real masterpiece. It is an elegant composition, a frontispiece visible from the Boulevard Poissonnière, from the top of rue Rougemont.

What particularly caught my attention, the first time that I contemplated this magnificent fresco, were the two proues of ship, flanked on both sides of the frontispiece. Each nave is actually made up of three distinct naves, enchanted one in the other, and this appears more clearly when you look at the profile facade, the triple nave emerging in an elegant protrusion, seems to float on a garland of fruits. To increase the intrigue aroused by this motif, a caduceus is part of the axis of the bow, as a Masaine mast. Finally, the last detail, the figurehead is a jellyfish head.

What a subtle way of representing the triple alchemical vessel associated with the subject of philosophy.

In the work of the President of Spagnet, one can read at the hundred ninth canon:

"The vase in which philosophers cook their work is of double origin; One is of nature, the other is art. The vase of nature, which is also called the vase of philosophy, is the land of stone, or the female, or the

First work: the rooster, the serpent, the mirror and the griffins

matrix in which the seed of the male is received, puts and is prepared for generation. Now the vessel of art is triple, indeed, the secret is cooked in a triple ship."

This vessel, our Athanor, is our matrix, it also represents the assembly of materials, mercury, sulfur and salt that the sage must elect to accomplish the work. It also represents the composition of the compost of the final coction. It is associated here with the caduceus, representing the mercury of the wise, recognizable by the two intertwined snakes and surmounted by a pair of wings. Curious this ship, which floats on a phylactère made up of a braid, abundantly supplied with fruits, vegetables and cereals of all kinds, an undeniable sign of abundance.

The last refinement, the figurehead is one of the gorgonians, the jellyfish, the same which also appears on the torso of Hermès and which, according to Fulcanelli, personalizes both prudence and wisdom.

In Greek mythology, a jellyfish, one of the three Gorgones, had trade with the God Neptune in the Minerva temple. The goddess, indignant at the desecration of her temple, then changed the jellyfish hair and gave her the ability to petrify all those she looked at. Perseus with the help of a magic shield knew how to look away and cut his head. Medusa blood was born Chrysaor and Pegasus. This legend relates to the three works of which it constitutes an abbreviated, Chrysaor evoking the chrysopée.

The central room of the frontispiece is the effigy of the god Mercury-Hermès, thriving majestically framed by two robust griffins, on a lintel, above the three doors of the building. Our illustrious character, in a majestic attitude, calmly sports a scepter surmounted by a bronze rooster in his dexter, and a mirror decorated with a snake in his senestre, while he lets rest his forearm on the mane of the griffon . The rooster also symbolizes philosophical mercury, and its presence on this scene is not accidental.

The two griffins of this composition have a lion's head, which is unusual. Conventionally the imagiers sculpt them an eagle's head and a fawn body with wings. The lion's head here marks power, the male principle, and the wings volatility. The assembly gives a hybrid,

half-fixed, half-voltile being, the griffin of the first work.

Other details surprise. The messenger also carries his bust with the head of Medusa, intertwined with snakes, and two merelles adorn the throne. This repetition of the pattern of the jellyfish must hide an important secret of the practice, that of the final coagulation of the stone.

The merelles, or scallops, above each griffin, attest to the philosopher's quality of the mercury obtained by the canonical route and mean that the operator has chosen the right materials.

Let's go back to the two griffins.

For Fulcanelli, in the second volume of philosopher's houses, the griffin marks the result of the first operation:

"From the fight that the knight, or secret sulfur, book with arsenical sulfur of the old dragon, is born the astral, white, heavy stone, brilliant as pure silver, and which appears signed, carrying the imprint of its nobility, the claw ..."

Further on, the master indicates the good proportions:

"If therefore, you want to have the griffin, - which is our astral stone, - by tearing it away from its arsenical gangue, take two virgin earth shares, our scaly dragon, and one of the igne agent, who is this valiant Knight armed with the lance and the shield... ".

We already know that the scaly dragon is the Stibin and the knight the iron. In alchemical cryptology, Stibin had the Crucifer globe symbol, this same globe appearing on all painted representations as well as statues of popes or sovereigns.

At the end of the alchemical combat that the old dragon and the martial knight engagement, Bernard Husson explains an operating detail never mentioned before, appearing on the illustration of the fourth key of Basile Valentin, the wheel of thea Fortune.

"The cruciferous globe unambiguously designated the trisulfure of antimony, while the antimony itself being its regulating. The Globe Crucifre overturned or returned identifies with the sign of Venus. Basile Valentin Apparent Venus with the Roman goddess Fortuna, standing on a globe, arms apart to keep her balance, thus expressing the precariousness of fortune. The alchemists of the 17th century knew this sphere well, whose smooth surface slides between the fingers of the artist because she is humid of lunar tears, which exert on the fingers a slightly caustic action ..."

Above all, it is necessary to retain the caustic qualifier, a term that can no longer be appropriate, expressing the action of potassium oxide partially transformed into hydroxide (koh), and which we find on the surface of this globe at the end of the fight.

He was simply generous, Bernard Husson. No one can no longer doubt now that he personally experiences the phase described above.

The operation of the completed griffon, the alchemist must undertake the reversal of the world (hear the upheaval or rather the reversal of the ball), which justifies the use of the symbol of Venus, which should not be taken for the vulgar copper.

Once the fight of the two protagonists are completed, it is simply necessary to overthrow the crucible in an iron horn, previously greased. Then the artist will operate the separation by hitting a dry blow on the cold mold to collect the vile trio. He will finally be able to open the door to the second work.

Another interesting practical detail is on the composition, the snake mirror. The mirror is the hieroglyph of universal matter. Mirror and subject of wise are synonymous in alchemical language. But the symbolism is complicated because there is ambivalence, it is a rule in alchemy. The mirror can represent both sulfur and mercury, or both, simultaneously, in the emblematic character of the rebis.

For the cosmopolitan, speaking of sulfur: *"In his kingdom, there is a mirror in which we see everyone. Anyone who looks in this mirror, can see and learn the three parts of everyone's sapience, and in this way, he will become very learned in these three reigns, as Aristotle, Avicenna, and several others were."*

For Basile Valentin: *"The entire body of vitriol must be recognized for a mirror of philosophical science."*

But the mirror of art is also, in its concavity, the emblem of reflective, androgynous mercury containing both the agent and the patient, which indicates the serpent carved on his reverse.

In ornament, on both sides of the vault, there are symmetrically two mortuary crowns, placed on a laurel branch, stressing the importance of Caput Mortum in the work. The two mortuary crowns inform us that there are, in the philosopher's development, two operations each resulting in Caput Mortum:

– The first, in the Griffon operation, where it is above all should not reject the slag.

– The second Caput Mortum appears at the end of the coction, when the happy artist receives the escarboucle in an opaque, rough and red setting, called the damned earth of the stone, residue useless that must be thrown.

Under the lintel of the composition, we find the pattern of the caduceus, repeated twice and appearing above the extreme doors.

At the same level, above the central door, there is a winged helmet, symbolizing the volatilization of the fixed but also the required proportions: a fixed part (the helmet) against two parts of the volatile, the wings.

Above, above our Hermès-Mercure hero and at the base of the pyramidal roof, the sculptor staged two characters.

The statue of the left is very similar to that of the justice of the tomb of François II, in Nantes, abundantly commented by Fulcanelli in the second volume of philosopher's houses.

Here it is veiled and without sword. The veil is put on purpose, to underline the esoteric nature of symbolism. The same teaching comes from the closed book being one with the scales, symbols of weights and raw matter.

On the coverage of the book we guess the registration large book, in capital letters, what could be more natural than a book of accounts for a banking organization?

The artist, pliers without laughing, wanted to steal the hieroglyph of the raw material or the crude mineral, the great book of nature, dispensor of secret dissolvant, a book closed par excellence, representing the raw material out of leaving out of The mine, which alchemical work strives to ennoble in order to obtain the living and open mercury, the balance indicating the need for respect for weights.

The right statue represents an unveiled young woman, tightening in her left hand the caduceus

The absence of a veil informs us of the change of state of the subject of art which, following a series of secret operations, is transformed into philosophical mercury.

The use of female characters is constant in the symbolism of mercury to underline the passive and volatile nature of this principle.

Fountain of March, Paris, rue Saint-Dominique

La Fontaine de Mars

There is, on rue Saint-Dominique, a stone's throw from the Champs de Mars in Paris, a magnificent fountain whose waters are apparently excellent, so much so that many fans of the Paris comes there daily to stock up.

This small monument is one of the many curiosities of our capital, but the motifs of the facade, too discreet, apparently did not intrigue ordinary people.

The Fountain of Mars, erected under the Empire, at 188 rue Saint-Dominique, near the Eiffel Tower, was also called the Fontaine de l'Hospice Militaire du Gros Caillou.

On the front side, the sculptor has depicted a couple representing Mars and Venus in full embrace. The god of war, practically naked, is wearing his helmet. In his left hand he clutches a sword in its sheath, while his right arm seems to rest on a shield located in the background.

Venus, in a vaporous dress, her hair adorned with a laurel wreath, holds a cup in her right hand. A snake coils around his right forearm.

Two other details attract the attention of the lover of old stones: the wave that is discreetly looming behind our two characters and which seems to come from a tower in the background (it could just as well succeed), and a rooster which stands on the shield, at the feet of March.

As a complementary ornament, we will notice, each floating on the wave, two fabulous animals, the pattern repeating on the four sides of the base of the building. These two animals have a fish body, but the one on the left has a lion's head while the one on the right has a ram's head.

Finally, on each side, on both sides of the front of the monument, there is a relief urn on which three characters appear: a man holding a disc in each hand, a woman who runs with a budding branch hand and a gnome, brandishing an instrument seeming to be a tambourine, with the right hand, and holding in his left hand, a small budding branch.

The rear facade is naked which has something to amaze.
The reader will not be surprised to learn that this magnificent fountain is entirely dedicated to alchemy.

Let us first examine the characters of the facade.
The Mars-who has come to the materials of the great work. Mars, planet representing iron, symbolizes the active, male agent. The woman that we could associate with Venus, but which should not be confused with copper, symbolizes the female and passive subject of art. In order for there to be no confusion, the scaplter has included a snake around the arm of Venus, which is devoted to the god Mercury, a female principle. The rooster, placed under the protection of Mars, underlines the esoteric character of the composition. The artist who designed this work has multiplied all the details allowing the indisputable interpretation of the airtight motif, making one last detail, attesting to the quality of the female agent: the cut that Venus holds in his right hand and that She seems to offer the Mars God. The cut symbolizes our vase, and this composition relates to the first work.
Our Venus, slightly dressed, is the mineral having undergone a first preparation before the nuptial ceremony, and Mars, practically naked, the valiant knight many times mentioned. Bernard Husson has shown that the clothes of the characters appearing on alchemical scenes are, most of the time, in relation to the operating phase and his progress. The quadling of the shield is evocative of the sizzle which signs the pefection of the compound as well as the Vulcan net.
The small column that we see slightly protruding, bottom left of the composition, just behind Venus, looks like a tower towards which the almost erased wave seems to be directing, or else, another event, in an oven of where fumes emerge. In the first hypothesis, this representation is to be compared to the fresco of the cathedral of Amiens, representing the dew of the philosophers, falling on the secret oven.

Mars fountain - Close-up of the front panel

In the second possibility, which seems most likely, the tower relates to our active Athanor, as evidenced by a relatively supplied smoke that escapes it.

The fabulous figurines swimming on the waves are emblematic of the result obtained at the end of the eagles or sublimations of the second work.

Indeed, it is the result of the first coagulation of mercurial water that the follower Fulcanelli, commenting on page 33, volume II of philosopher's residences, describes as mysterious, "*Both by its development contrary to the chemical laws as by its Obscure mechanism, mystery that the best educated scientist and the most expert follower cannot explain.*"

These marine monsters offer the same meaning as the dolphin or remained; They symbolize mercury resulting from philosophical sublimations. The scales of the body of fish floating on the waves appear the intertwined lines which appear on the properly prepared matter, just like this braid or fishnet, which appears a few times on esoteric representations, assimilated to the Vulcan net which surprised Mars and Venus in Alfutère, and that we noted on the shield.

The heads of lion and ram relate to the essential condition of start-up, from the start of the work, "*this is one of the greatest secrets of work, and, whatever it is, stone stumbling blocks on which the too much of the researcher is breaking, from the threshold,* "warns Claude d'Ygé in his new assembly.

For Fulcanelli, "*This is the most delicate phase of work than that where the coagulation of the stone, creamy and light, seems on the surface and floats on the waters. It is then necessary to redouble a precaution and caution in the application of theFire, if you don't want to blush it before and rush it. She manifested herself at the beginning in the appearance of a thin film, very quickly broken, whose detached fragments of edges retract, then weld, thickens, take the form of a flat island, - the island of the cosmopolitan and the mythical earth of delos, - animated by roundabout movements and subjected to continual trips. This island is only another figure of the hermetic fish, born from the Sea of Sages, - our Mecure that Hermès calls Mare Patens, - the pilot of the work, the first solid state of the embryonic stone.*"

The symbolism of side sculptures is less transparent. The relief athlete that appears on the urn, holds a disc in each hand. The right disc is pierced in its center, it is the symbol of gold or hermetic sun.

The second disc, largely masked by the left hand, corresponds to the lunar hieroglyph. These two discs have the same esoteric meaning as those of the Statue of Saint-Marcel terraging the Dragon, of which Fulcanelli made a masterful demonstration in the mystery of cathedrals. The discs here evoke open materials and ready to deliver their active ingredient. We will also find a reason for similar meaning on the handles of the urn, in the form of rosettes. In the latter case, these are the two exalted principles, sulfur and mercury philosophical.

The Venus of the Urn is revealed. The allegory of this table is a lightened version of that appearing on the engraving illustrating the second Valentin Basile key:

"A virgin, having to be given in marriage, is first of all beautifully adorned with a variety of the most precious clothes, so that it pleases its fiancé and that, by its appearance, it lights in it, deeply, the 'Hit of love. But when she must be married to her fiancé, according to the use of the carnal union, it is removed all her different clothes and she keeps none, if not the one given to her by the Creator , at the time of his birth."

Although the bishop blessing the Magic Union of Mars and Venus is absent from the scene, the Imagier did not, however, omitted the Salin Mediator in this composition. Better still, it makes it include, in place of the bishop, a small monstrous character whose posterior part resembles the lower limbs of a goat or a goat. Could it be a gnome?

The sponsor of this work therefore knew perfectly the symbolism and the scope of this operation, one of the most secret in the process.

Fulcanelli, in his analysis of the solar dial of the Hollyrood palace in Edingburg, calls on the language of birds. The Greek root of the word gnome, he specifies, means spirit, intelligence, and it continues:

"Now, the Gnomes, underground geniuses attended by the care of mineral treasures, constantly watch over the gold and silver mines, the lodgings of precious stones, appear as symbolic representatives, humanized figures of the Vital Spirit metallic and material activity.

Tradition portrays them as for ugly and very small stature; On the other hand, their naturalness is gentle, their carceries beneficial, their trade extremely favorable."

Further on, completing the analysis of the Arcane, evoking the Sun Marche of the Sages in the work of philosophy, he adds:

"And this march is set by the icosahedron, which is this unknown crystal, the salt of sapience, spirit or embodied fire, the familiar and helpful gnome, friend of the good artists, who assures man the accession to the supreme knowledge ancient gnosis."

There is no doubt that this is the same gnome here, symbolizing the salt of Sapience, the emerald of philosophers, that is to say the secret catalyst, our Vytriol.

There is still a detail that may deserve to be emphasized: the gnome seems to make music with his tambourine which he brandishes with the right hand. Alchemy, in fact, is synonymous with music art for large number of artists who have achieved the wonderful harmony. It is a discreet allusion to certain sound signs of the coction.

Regarding the budding branches that our female character and the gnome hold, it is an allusion to the gem that the happy artist can contemplate at the end of the third work. The word Bourgeon comes from the Latin Burrionem, accusative of Burrio, Burra, which means stuffing, gem, pushes or rejection, which is not in contradiction with our subject. Gem or young shoot, isn't chemistry considered to be celestial agriculture?

Finally, one of the leafy branches of the composition wraps, like a snake, around the branch of our goddess. The Imagier probably wanted to discreetly emphasize the mercurial character of the goddess.

As for the urn, it again symbolizes our vase, the secret vessel, receptacle of philosopher's water, a land of the poor man's way, perfectly suited to Brevi art.

Mars fountain - Detail of the lateral motif

Addendum

The Wet Way

Cyliani, in the Preface to his book, Hermes Unveiled, describes the two processes, dry and wet, in these terms:

"I believe I am warning here to never forget that we only need two materials of the same origin, one volatile, the other fixed; that there are two ways, the dry way and the wet way. I follow the latter, preferably, out of duty, although the first is very familiar to me: it is made with a unique material."

According to Fulcanelli, the two paths, dry and wet, begin with the same operation and it would be from reaching the Caput that the two paths branch off.

Indeed, Cyliani very charitably tells us the process that he involves just after the fight between the alchemist and the dragon of the first operation, that is to say the reduction of the stibnite by iron in a crucible over open fire.

In all likelihood, the artist must recover the slag and abandon the metal.

This process has been described by several alchemists including the anonymous author of *Hermetic Recreations*.

The process consists of spreading the supply of slag recovered from the 3 reiterations of the first work to expose them on an inclined dish placed above a vase fitted with a funnel, at night, in a cellar, in order to recover the liquid that the hygroscopic slag will inevitably produce by absorbing moisture from the air

This is also the reason why the author of Hermetic Recreations recommends choosing the right period. He warns against the temptation to use dew or rainwater

"I have already made it known that it was not rain water or dew water which was suitable for this operation, I will add here that it is neither the water from a species of fungus commonly called Flos Coeli or Flower of the Sky and which we very incorrectly take for the Nostoch

of the ancients, but an admirable water drawn by artifice from the rays of the sun and the moon I will also say that the salts and other magnets which we use for. extracting moisture from the air, are good for nothing in this circumstance and that there is only the only fire of Nature which we can use usefully here. This fire contained in the center of all bodies needs. of a certain movement to acquire this attractive and universal property which is so necessary to you, and there is only one body in the world where it is found with this condition, but it is so common that we encounter it wherever man can go; therefore I think it will not be difficult for you to meet him."

Further he specifies the ideal period:

"But there are still certain essential provisions to be fulfilled, without which you would only have clear and useless water.

There is only one time suitable for this harvest of astral spirits. It is the one where Nature regenerates; because at this time the atmosphere is completely filled with the universal spirit. The trees and plants that turn green again, and the animals that indulge the pressing need of generation, make us particularly aware of its benign influence. Spring and autumn are therefore the seasons you should choose for this work; but spring especially is preferable. Summer, because of the excessive heat which expands and chases away this spirit, and winter because of the cold which retains it and prevents it from exhaling, are out of order. In the south of France work can be started in March and resumed in September; but in Paris and in the rest of the kingdom, it is only in April that it can be started and the second sap is so weak that it would be a waste of time to take care of it in the autumn."

And for good reason, the slag must be able to attract humidity from the air and avoid the harmful action of UV rays, hence the need to harvest at night in spring or autumn. The process perfectly described by Cyliani which involves a Nymph, symbol of the liquid fraction collected at night, consists of a series of imbibitions and deications.

It is necessary to collect a very large quantity of liquid for the tedious series of imbition. You need plenty of distilled spirit water, recommends Arnaud de Villeneuve

We see that we should not use a closed vessel as this would make the operation practically impossible. Glass Petri dishes should be used.

The simple process will consist of soaking the slag with the liquid collected throughout the cooking, which will last approximately two years, without drowning the substrate which would cause its ruin. Then we leave the compost to dry out, always away from light, and we will continue imbibing until the substrate is again saturated with its humidity.

Here is what the anonymous author of les *Récréations Hermétiques* says:

"The vessel of Nature is the prepared earth that must be watered with one's spirit. It is called a vessel, and indeed it is, since it contains. The spirit that we add to it is not a foreign thing since everything came out of it, and our earth is formed from it; this is why it is said to bring the child into the mother's womb: which can only be done by tearing out its entrails. Our earth must also be divided into its smallest parts to bring its great riches to light, and this will happen if you often fill it with its spirit and let it dry out as many times. In this operation, the phlegm evaporates, but the spirit remains and is incorporated with the earth which it salifies until saturation is complete; then the spirit that is added, no longer able to be contained, reacts on that which the earth has fixed and forces it to dissolve, as would salt; this is why this dissolution is compared to a sea; and because the spirit that is added is joined to an altering and corrupting humidity, its mixture results in a movement of fermentation which is followed by putrefaction, and consequently by regeneration, because fermentation changes the bodies of Nature, and in putrefaction, they only exchange their clothes for new ones, all the richer and more brilliant, as the driving Spirit is of a higher origin.

The amount of moisture Matter can contain, without releasing it outside, is the measure to observe for imbibitions, and what we call the weight of Nature.

The matter serving as a vessel also serves as a furnace, since the spirit that you introduce into it is a natural fire which cooks and digests it to serve me, until the end, philosophical expressions.

No less than fifty ablutions are required; because each ablution until perfect desiccation is counted as a natural or philosophical day; so that our days can last a week depending on the season, the quality and the quantity of material submitted to the work. The great secret of the Sages to shorten time is to divide matter so that the days have less length."

Further away:

"When we have done this and the material dissolves, it gradually blackens. In these various times only the spirit necessary to maintain its fermentative fire is added; and when the matter begins to ferment, it must be left to its own fire, until the perfect whiteness where it arrives of its own accord.

The material is not liquid like broth, but thick and black like pitch or boot polish; it swells, rises in the Goblet, gives Bubbles which are compared to the eyes of a fish and which must not be burst, because they contain the animating spirit.

After fermentation, the material collapses; it is then shiny like pitch, and the most beautiful black; this is the sign of putrefaction which we call crow's head. It then dries out little by little and changes to an ash gray color. Soon a capillary circle of the most dazzling whiteness appears around the vessel. This Circle widens more and more until everything is perfectly white."

The sequence of operations being perfectly described by Cyliani in Hermes Dévoilé, and in the sRécrétions Hermétiques, I will not dwell too much on the wet process.

Just a note about using commercial mercury Hg.

Many alchemists such as Flamel, Urbiger, Philalethe use hydrargyrum at a certain stage of the work, an operation which does not have the consent of Fulcanelli, but it is what Cyliani does in the third work in order, he says , to shorten the work

You will therefore find the sequence of operations perfectly expliained in the two best works mentioned above, Hermes Unveiled and the Récrétions Hermétiques.

I hope you were not bored reading my book
Thank you for your interest.

Bibliographic index

BARBAULT, Armand, *L'or du millième matin,* Paris, J'ai Lu, 1970

BARCHUSEN, Johann Conrad, *Figures hermétiques,* in Alchimie les cahiers de l'hemétisme, Paris, Éd. Dervy, 1996

BADEAU, Fabrice, *Les clefs secrètes de la chimie des anciens,* Paris, Robert Laffont, 1975

BARRENT COENDERS Van Helpen, The staircase of the wise, Milan, Arché, 1976

BASILE VALENTIN,

• *The Twelve Keys of Philosophy*, translation and commentaries by Eugène Canseliet, Paris, Éd. de Minuit, 1968

• *The Triumphal Chariot of Antimony,* introduction by Sylvain Matton, Paris, Retz, 1977

•*The Last Testament*, introduction by Sylvain Matton, Paris, Retz, 1978

• Revelation of the Mysteries of the Tinctures of the Seven Metals, Paris, Psyché, 1954

BATFROI, Séverin, *Alchemical metamorphosis of universal mercury*, Paris, Guy Trédaniel, 1976

BELIN, Dom Jean-Albert, *The adventures of the unknown philosopher* by Sylvain Matton, Paris, Retz, 1976

BONARDEL, Françoise, *Philosophe par le feu,* Paris, Éd. du Seuil, 1995

BURCKHARDT, Titus, *Alchemy, science and wisdom.* ,Paris, Éd. Encyclopédie Planète, 1967

BURLAND, Cottie Arthur, *Le savoir caché des alchimistes.* Paris, Robert Laffont, 1969

CAMBRIEL, Louis-Paul-françois, Cours de philosophie hermétique en 19 leçons.

CANSELIET, Eugène,

• *Deux logis alchimiques*, Paris, Éd. Jean Schemit, 1945

• Alchimie, études diverses de symbolisme hermétique et de pratique philosophale. Paris, J.-J. Pauvert 1964

• *L'Alchimie et son livre muet, Mutus Liber.* Paris, J.-J. Pauvert 1967

- *Trois anciens traités d'Alchimie*. Paris, J.-J. Pauvert 1975
- *L'Alchimie expliquée sur ses textes classiques*. Paris, J.-J. Pauvert 1980

LE COSMOPOLITE, *Nouvelle lumière chymique.*, Paris, Retz, 1976

CYLIANI, Hermès dévoilé-See HUSSON Bernard.

ECKARTSHAUSEN, Karl von, *Essais Chimiques*. Paris. Psyché, 1963.

ESPAGNET, Jean d', L'*Œuvre secret de la philosophie d'Hermès*.,Paris, Denoël, 1972
Excellent classic. The theory is boring, however, the practice developed is very rich.

ETTEILA, *Les sept nuances de l'œuvre*, Arma Artis 1977.

VADIS, Egidius de, *Dialogue entre la Nature et le fils de la Philosophie*. Paris. Éd. Dervy, 1993

EYQUEM de MARINEAU, Mathurin, *le Pilote de l'onde vive*. Paris, E..P. Denoël, 1972

FABRE, Pierre-Jean, *Abrégés des secrets chymiques,* Paris, Gutenberg Reprint, 1980

FIGUIER, Louis, *L'Alchimie et les alchimistes*. Paris. Éd. Denoël, 1970

FLAMEL, Nicolas
- *Le livre des figures hiéroglyphiques, Le Sommaire, Le Désir désiré* Paris, S.G.P.P. Denoël, 1970
- *Les œuvres de Nicolas Flamel, contient Le livre des figures hiéroglyphiques, Le Sommaire, Le livre des laveures, le Bréviaire de Nicolas Flamel*. Paris, Pierre Belfond 1973

FULCANELLI
- *Le Mystère de Cathédrales et l'interprétation des symboles herméthiques du Grand Œuvre*. Paris, J.-J. Pauvert 1964
- *Les Demeures Philosophales et le symbolisme hermétique dans ses rapports avec l'art sacré et l'ésotérisme du Grand Œuvre*. Paris, Jean-Jacques Pauvert 1965

GEBER, *La somme de la perfection*. Paris, Guy Trédaniel, 1976

GLASER, Christophe, *Traité de la chimie. Paris,* Gutenberg Reprint, 1980

GOBINEAU DE MONTLUISANT, t*rois traités d'Alchimie*

GRASSOT, Louis, *La lumière tirée du chaos. Paris,* Gutenberg Reprint, 1980

GRILLOT DE GIVRY, Émile-Jules,
- *Lourdes, ville initiatique*. Paris, Éd. Traditionnelles 1979
- Le Grand Œuvre. Paris, Éd. Traditionnelles 1981

HERMÈS TRIMÉGISTE, *La table d'Émeraude et sa tradition alchimique*. Paris,

Belles Lettres, 1994, Content:
- Le livre du Secret de la création d'Appolonius deTyane.
- La Table d'Émeraude.
- Le livre des sept traités d'Hermès Trimégiste.
- Le livre de Crates.

HUGINUS À BARMÂ, *Le Règne de Saturne changé en siècle d'Or*, followed by *La Pierre de Touche*. Paris, Éd. Pierre Derieu, 1780

HUSSON, Bernard,
- *Deux anciens traités d'alchimie du XIXème siècle*. Paris, Éd. Omnium Littéraire, 1964
 Contains the treatises below:
- CAMBIEL, Louis-Paul-François, *Cours de Philosophie hermétique en 19 leçons*, 1843.
- CYLIANI, *Hermès Dévoilé*, 1832.
- ANONYMOUS, *Récréations Hermétiques*.
- *Anthologie de l'Alchimie*. Paris, Pierre Belfond 1971

This work contains in particular:
- FRANÇOIS, René, *Essais sur la Rosée*. Paris. 1621.
- ANONYMOUS, *La parabole de Mars de Busto Nicenas*. 1619.
- TOLL, Jacques, *Le chemin du Ciel Chimique*. Amsterdam, 1688.

URBIGER, *Aphorismes*. Hambourg. 1705.
- Transmutations Alchimiques. Paris. Éd. J'ai lu 1974, 1 Vol.
- *Trois Textes inédits du XVIIème siècle*. Paris, Librairie de Médicis. 1979

This work contains
- *L'or potable des anciens*.
- *Lettres philosophiques*
- *Le testament d'Or*
- *Viridarium Chimicum*, Paris, Librairie de Médicis, 1975

KHUNRATH, Henri, A*mphithéâtre de l'éternelle sagesse*. Lyon, Paul Derain, 1957
LAMBSPRINK, *Traité de la Pierre Philosophale*. Paris, E.P. Denoël, 1972
This work contains:
- *Le Pilote de l'Onde vive, du Sieur Mathurin Eyquem*.

LE BRETON, *Les Clefs de la philosophie spagyrique*. Paris, Gutenberg Reprint, 1976
LE TRÉVISAN, Bernard. Paris, Trédaniel-La Mesnie, 1991 - This work contains
- *Le livre de la philosophie naturelle des métaux,*

- La parole délaissée,
- *Le songe vert,*
- *Le traité de la nature de l'Œuf philosophique.*

LIMOJON DE SAINT-DIDIER, Alexandre Toussain, *Traité de la Pierre Philosophale.* Paris, E.P. Denoël, 1971

This work contains:
- *Le Mutus Liber, hypotypose de Magophon,*
- *Le Triomphe hermétique*

LINTHAUT, Henri de, *L'Aurore -d l'Ami de l'Aurore.* Paris, Trédaniel-La Mesnie, 1978

ANONYMOUS
- La Lumière sortant par soi-même des ténèbres. Paris. E.P. Denoël, 1971 1 Vol. Introduction et notes de Bernard Roger.

This work contains:
- *Aphorismes chimiques ou véritable théorie de la Pierre,* Bruno de Lanzac.
- *Le Livre de Senior, Lettre de Psellos sur la Chrysopée.* Paris, Éd. Dervy, 1993

LEO, Irénéus, *Compendium des trois Œuvres Chymiques. - Initiation à la voie de l'Oratoire.* Paris, Éd. Ramuel, 2003
- *Le Rosaire des Philosophes.* Paris, Librairie de Médicis, 1973
- *La Table d'Émeraude,* See HERMÈS TRIMÉGISTE.
- *Traité sur la Matière des Philosophes en général.* Paris, Trédaniel- La Mesnie, 1983.
- *La Tourbe des philosophes.* Paris, Dervy 1993
- *La Pierre Aqueuse de sagesse or l'Aquarium des Sages.* Paris. La Table d'Émeraude, 1989

MARCUS DE VEZE, Jean, *L'Or Alchimique.* Lyon, Éd. du Cosmogone 2001

MAIER, Michel, *Atalante Fugitive.* Paris, Éd. Dervy 1997

MONTE SNYDERS, Johannes de, *Commentaire sur la Médecine Universelle* Arché, 1977

MORAS DE RESPOUR, *Rares expériences sur l'Esprit minéral,* Paris. Jobert, 1975

PARACELSE, *Quatre traités. Paris,* Éd. Dervy, 1992

This work contains:
- *Le Labyrinthe des Médecins errants,*
- *Cinq traités de Philosophie,*

- *Le Livre de la restauration,*
- *Le Livre de la longue vie.*

PERNETY, Dom Antoine Joseph,
- *Dictionnaire Mytho-Hermétique*, Paris E.P. Denoël 1972
- *Les Fables Égyptiennes et Grecques dévoilées,* Paris, La Table d'Émeraude, 1981

PHILALÈTHE, Eyrénée,
- *L'Entrée ouverte au Palais fermé du Roi.* Paris. S.G.P.P. Denoël, 1970 .
- *Expérience sur la préparation du Mercure des Sages pour la Pierre par le Régule de Mars ou Fer tenant de l'Antimoine étoilé et par la Lune ou l'Argent.*
- *Règles du Philalèthe pour se conduire dans l'œuvre hermétique.* Gènes, Phœnix, 1979

PICCOLPASSI, Cyprian, *Les Trois livres de l'art du potier*, Paris, Éd. Ramuel.

PLANYS-CAMPY, David de, *Ouverture de l'Escholle de philosophie transmutatoire*, Paris. Guttenberg Reprint, fac similé 1980 .

PONTANUS, Jean, *Épitre du feu philosophique*. Paris. Guy Trédaniel-La Mesnie, 1981

ANONYMOUS : *Le Psautier d'Hermophiles envoyé à Philalèthe.*Paris. Éd. Dervy, 1997

RANQUE, Georges, *La Pierre Philosophale*. Paris. Robert Laffont. 1972,
This work contain:
- LAMBSPRINCK, *Traité de la pierre philosophale,*
- BASILE VALENTIN, *Des choses naturelles et surnaturelles,*
- LIMONJON de Saint-Didier, *le Triomphe Hermétique.*

RIPLEY, George, *Les douze portes de l'Alchimie*, Paris. Trédaniel-La Mesnie, 1979

RICORDEAU, Joseph, *l'Œuvre au Blanc*, Paris. Éd. Traditionnelles, 1975

RIVIERE, Patrick, *Alchimie et Spagyrie*, Éd. de Neustrie, 1988
- *La Médecine de Paracelse*, Paris, Éd. Traditionnelle 1988
- *L'Alchimie science et mystique,* Paris Éd. de Vecchi 1990
- *Fulcanelli, sa véritable identité révélée. Paris.* Éd. de Vecchi
- *Alchimie et Archimie, l'art des particuliers,* Paris Éd. de Vecchi

SADOUL, Jacques, *Le trésor des alchimistes*. Paris Denoël 1970

STOLCIUS, Daniel, *Viridarium Chymicum*, Paris. Librairie de Médicis, 1975

STUART DE CHEVALIER, Sabine, *Discours Philosophiques sur les trois principes, animal, végétal et minéra*l, Paris. Gutenberg Repint, 1982

TOLLIUS, Jacques, *Le chemin du Ciel Chimique.* Amsterdam. 1688,

TRÉVISAN, Bernard, *La parole délaissée,*

TRISMOSIN, Salomon, *La Toison d'or,* Paris, Retz, 1975

VALOIS, Nicolas et GROSPARMY, Paris, Retz, 1975

This work contains/

- *Le Trésor des Trésor,*
- *Les 5 Livres - la Clef du Secret des Secrets.*

VEZE, Marcus, *L'Or Alchimique,* Lyon, Éd. du Cosmogone, 2001

VIGENÈRE, Blaise de, *Traité du Feu et du Sel.* Paris, Jobert, 1976

D'YGÉ, Claude, *Nouvelle assemblée des philosophes chymiques,* Paris, Dervy-Livres, 1954 - This work contains:

- *Extrait des 5 livres de Nicolas Valois,*
- *Extrait des Aventures du Philosophe Inconnu, de Belin,*
- *Science écrite de tout l'art hermétique* (anonymous),
- *Explication très curieuse des énigmes et figures hyéroglyphiques de Gobineau de Monluisant,*
- *La parole délaissée,* by Bernard le Trévisan.

www.ingramcontent.com/pod-product-compliance
Lightning Source LLC
Chambersburg PA
CBHW062313220526
45479CB00004B/1152